Emerging Frontiers in Systems and Synthetic Biology 2015

AA003257

Topical Conference at the 2015 AIChE Annual Meeting

Salt Lake City, Utah, USA
8-13 November 2015

ISBN: 978-1-5108-1847-7

Printed from e-media with permission by:

Curran Associates, Inc.
57 Morehouse Lane
Red Hook, NY 12571

Some format issues inherent in the e-media version may also appear in this print version.

Printed by Curran Associates, Inc. (2016)

For permission requests, please contact AIChE
at the address below.

AIChE
120 Wall Street, FL 23
New York, NY 10005-4020

Phone: (800) 242-4363
Fax: (203) 775-5177

www.aiche.org

Additional copies of this publication are available from:

Curran Associates, Inc.
57 Morehouse Lane
Red Hook, NY 12571 USA
Phone: 845-758-0400
Fax: 845-758-2634
Email: curran@proceedings.com
Web: www.proceedings.com

TABLE OF CONTENTS

422663 The Dionesus Algorithm Provides Scalable and Accurate Reconstruction of Biological Networks to Reveal New Drug Targets

Abstract:

The DIONESUS algorithm provides scalable and accurate reconstruction of biological networks to reveal new drug targets.

Mark F. Ciaccio and Neda Bagheri. Department of Chemical & Biological Engineering, McCormick School of Engineering, Northwestern University.

Introduction: Advances in genomic and proteomic technology brings with it an unprecedented breadth of data on the molecular markers driving cellular function. With this new age in biomedical engineering comes the pressing need to find efficient strategies to glean deeper mechanistic understanding of biological processes. Scalable and rapid algorithms are critical in pushing forward innovation at the interface of medicine, mathematics, and biology. To this end, we created a scalable algorithm, termed DIONESUS, based on partial least squares regression with variable selection to reconstruct networks from large matrices of cue-signal-response relationships. Our scalable algorithm performs with higher specificity and sensitivity than several top-performing network inference algorithms in a fraction of the computational time. We applied DIONESUS to reconstruct the signaling network of a model epidermoid cancer cell line and identified STAT1 as a central modulator of apoptosis specifically in cells that have hyperactivity of the EGF receptor. **Background:** Deriving correlations between biomolecules, such as RNA expression or protein abundance, and cell phenotype can suggest factors that are associated with disease. Deriving the underlying network structure can identify disease drivers by elucidating control structures such as feedback loops and redundant pathways. Several algorithms can reconstruct directed biological networks using information theory, decision trees, or Bayesian networking; algorithms with high accuracy, however, are often limited to small networks [1]. Ideally, network reconstruction can be expanded to genomic or proteomic proportions in order to broaden and deepen understanding of the dysregulation of biological systems in pathological states. **Materials and Methods:** After being validated in three *in silico* networks, DIONESUS was applied to a high confidence dataset characterizing 60 phospho-proteins at four time points as a function of 26 perturbations. We perturbed epidermoid carcinoma cells with diverse growth factors and/or small molecules chosen to activate or inhibit specific subsets of receptor tyrosine kinases. The abundance of 60 phosphosites was quantified using a modified microwestern array, a high-confidence assay of protein abundance and modification [2]. **Results and Discussion:** From the reconstructed network, we identified enhancement of STAT1 activity as a potential strategy to treat EGFR-hyperactive cancers. Quantification of the relationship between drug dosage and cell viability in a panel of triple-negative breast cancer cell lines validated this newly proposed therapy. **Conclusions:** The accurate and scalable DIONESUS algorithm allows for quick iteration between experimental design and computational modeling by incorporating systems-level behavior in order to accelerate drug target discovery.

References:

[1] Marbach, Daniel, et al. "Wisdom of crowds for robust gene network inference." *Nat meth* 9.8 (2012): 796-804.

[2] Ciaccio, Mark F., et al. "Systems analysis of EGF receptor signaling dynamics with microwestern arrays." *Nat meth* 7.2 (2010): 148-155.

429129 An Integrated Engineering Control System Model of Primary and Secondary Hemostasis

Abstract:

Hemostasis, the physiological process that arrests bleeding and keeps blood within an injured blood vessel, consists of a set of reactions and sub-processes that can be divided into various functional components. Normal functioning of the components is necessary to achieve a delicate balance between effective hemostasis and pathological conditions such as thrombosis, one of the major causes of death in the world, or hemophilia. Because the components of the hemostatic process interact in complex ways, how the various components work together to implement fast, effective and stable responses to vascular injury cannot be fully understood using traditional experimental methods alone—quantitative models are required. However, developing a faithful mathematical representation of known mechanistic details of the entire hemostatic process using standard modeling approaches will produce a model containing an inordinately large number of variables and an even larger number of unknown parameters. An alternative conceptual paradigm is required if the important complex details are to be represented adequately, with sufficient fidelity, and in mathematically tractable form.

From a process engineering perspective, we observe that hemostasis is mediated by an automatic biological control system with clearly identifiable control system components such as sensors, controllers, and actuators. Our modeling approach uses such an engineering control system representation as the basis for organizing information efficiently into a holistic model, with the control system components as "functional modules," thereby facilitating development, validation and systematic analysis of the complete hemostatic process as a combination of the component modules.

In this presentation, we will discuss the development of a comprehensive hemostasis model, and its validation against experimental data from the literature. We will first present a high-level engineering control system block diagram for hemostasis that shows the two major, distinct sub-processes: primary hemostasis (responsible for platelet aggregation) and secondary hemostasis (responsible for coagulation) and their interconnections. We will then present mathematical models developed and validated for each sub-process separately and subsequently connected via three known interaction mechanisms: (1) Platelet activation and aggregation (via primary hemostasis) provides biologically active surfaces for the enzyme reactions of the coagulation cascade (in secondary hemostasis); (2) thrombin (produced by secondary hemostasis) is a potent agonist for platelet activation (in primary hemostasis); and (3) platelet activation (in primary hemostasis) triggers the release of coagulation factor V stored within platelets, which in turn promotes thrombin production (in secondary hemostasis). We will then show results of the dynamic characteristics of hemostatic response predicted by our comprehensive model, and demonstrate how the interactions between the components in hemostasis contribute to the overall effectiveness of the hemostatic process.

The validated model will subsequently be used for systematic analysis of the overall integrated process of hemostasis and to generate hypotheses for rational diagnosis and treatment of pathological hemostasis conditions.

435404 Development of Kinetic Models of Metabolism for Microbial Production

Abstract:

Development of kinetic models of metabolism for microbial production

Ali Khodayari[1] (auk241@psu.edu), Satyakam Dash[1] (sud25@psu.edu), Matthew K. Theisen[3], M. Ahsanul Islam[2], Yuting Zheng[2], James C. Liao[3], Gregory Stephanopoulos[2], and **Costas D. Maranas**[1]

[1]The Pennsylvania State University, University Park; [2]Massachusetts Institute of Technology; [3]University of California, Los Angeles.

Improving the accuracy of *in silico* metabolic models is the chief challenge for accurately identifying the metabolic drivers that underpin the production of biofuels and chemicals of interest. Despite several advances, stoichiometry-based approaches are limited to steady-state conditions, and generally are unable to describe the metabolic phenotype in terms of the underlying pool of metabolite concentrations and enzyme abundances. While kinetic models have a potential to address these shortcomings, their constructions are still plagued by a number of challenges, including the paucity of kinetic parameters, diversity of rate laws, and implementation of regulatory events. The Ensemble Modeling (EM) paradigm [1] has addressed some of these challenges by decomposing reactions into their elementary steps thus enabling systematic integration of substrate-level regulatory interactions. Recently, we have constructed a kinetic model of *E. coli*core metabolism including 138 reactions, 93 metabolites, and 63 regulatory interactions under aerobic growth condition with glucose as the sole carbon source [2].

Here, we expand the core kinetic model towards genome-scale by integration of all relevant reactions from the *i*AF1260 model. Model parameterization is performed using experimental data for multiple carbon substrates (glucose, acetate, and pyruvate) under aerobic respiration and fermentative conditions. In addition, we developed a second generation genome scale model (GSM) for *C. thermocellum*(iCth446) which includes 446 genes, 598 metabolites, and 637 reactions. The GSM was subsequently used to construct a core metabolic model of the organism's central metabolism containing 90 reactions and 76 metabolites, with cellobiose as the carbon source under anaerobic growth condition. This core metabolic network spans all major biomass precursors, as well as other chemicals of interest encompassing the cellobiose degradation pathway, glycolysis/gluconeogenesis, the pentose phosphate (PP) pathway, the TCA cycle, major pyruvate metabolism and anaplerotic reactions, alternative carbon metabolism, nucleotide salvage pathway, and 22 substrate level regulatory interactions. The constructed core model serves as the scaffold for building a kinetic model parameterized by using experimentally measured flux data to improve isobutanol production.

1. Tan Y, Liao JC: **Metabolic ensemble modeling for strain engineers**. *Biotechnology journal* 2012, **7**(3):343-353.

2. Khodayari A, Zomorrodi AR, Liao JC, Maranas CD: **A kinetic model of Escherichia coli core metabolism satisfying multiple sets of mutant flux data**. *Metab Eng* 2014, **25**:50-62.

433386 GEM-Path: A Novel Pathway Prediction Algorithm and Commodity Chemical Production Atlas

Abstract:

The production of 75% of the current drug molecules and 35% of all chemicals could be achieved through bioprocessing. To accelerate the transition from a petroleum-based chemical industry to a sustainable bio-based industry, systems metabolic engineering has emerged to computationally design metabolic pathways for chemical production. Although algorithms able to provide specific metabolic interventions and heterologous production pathways are available, a systematic analysis for all possible production routes to commodity chemicals in *Escherichia coli* is lacking. Furthermore, a pathway prediction algorithm that combines direct integration of genome-scale models at each step of the search to reduce the search space does not exist. Previous work performed a model-driven evaluation of the growth-coupled production potential for *Escherichia coli* to produce multiple native compounds from different feedstocks. In this study, we extended this analysis for non-native compounds by using an integrated approach through heterologous pathway integration and growth-coupled metabolite production design. In addition to integration with genome-scale model integration, the GEM-Path algorithm developed in this work also contains a novel approach to address reaction promiscuity. In total, 245 unique synthetic pathways for 20 large volume compounds were predicted. Host metabolism with these synthetic pathways was then analyzed for feasible growth-coupled production and designs could be identified for 1271 of the 6615 conditions evaluated. This study characterizes the potential for *E. coli* to produce commodity chemicals, and outlines a generic strain design workflow to design production strains.

431865 Prediction of Gene Expression Using Goal-Directed Metabolic Models

Abstract:

The salient feature of cybernetic models, their goal-oriented control policies, have been used to predict metabolic phenomena ranging from complex substrate uptake patterns, and dynamic metabolic flux distributions to the behavior of gene knockout strains. The key element of this approach is the perspective that metabolic regulation is organized towards achieving goals relevant to an organism's survival. For example, in multi-substrate cultures of *E. coli*, cells invest resources into enzymes for pathways which maximize their growth rate. While goal-oriented control has successfully yielded the prediction of numerous metabolic phenomena, the control policy itself has not been compared with data for metabolic regulation. In this talk, cybernetic control variables are weighed against gene expression data, an indicator of cellular regulation, for two sets of systems.

The first set of systems analyzed consists of bacterial cultures that exhibit diauxic growth in two-substrate environments. For these bacterial models, enzymes are synthesized to maximize the growth rate for a given set of metabolic options at each point in time. Both microarray and qRT-PCR data for pathways pertaining to each substrate are compared with the cybernetic variable for enzyme formation. By demonstrating how cybernetic variables for induced enzyme synthesis mimic transcriptional data, a strong argument for using cybernetic models is made.

The second system modeled describes the behavior of a network of lipids in the mammalian macrophage cell line, RAW 264.7. More specifically, arachidonic acid is converted into various eicosanoids associated with inflammation in response to the detection of a key marker of infection, lipopolysaccharide, by Toll-like receptor 4 on macrophages. The existing framework for modeling single-celled organisms has been modified to describe more complex mammalian metabolic networks. Our approach essentially mimics, in an approximate sense, the maximization of the rate of TNF-alpha production, a crucial factor during inflammation, as an objective function. This is implemented by assuming that the upstream generation of eicosanoids concomitantly results in the downstream transcription of TNF-alpha. To be precise, enzyme synthesis for pathways that yield TNF-alpha production signals is regulated in such a way to ultimately maximize the signal for TNF-alpha formation. This is captured in the model using a linear equation that connects time-dependent changes in TNF-alpha with temporal changes in eicosanoid levels. Fitting this model to metabolite data and comparing cybernetic variables with dynamic gene expression data at the major branch point in this network shows a significant level of agreement. This result serves as an important validation of cybernetic variables using real data from cells. Moreover, it shows that cybernetic variables can be used to infer trends in gene expression data only using information taken from the metabolite level and a description of the metabolic network's organizing principle. This idea has not only been tested for a control condition and LPS treatment but now has also been tested on treatment with compactin, a statin drug that influences arachidonic acid metabolism.

419744 Network Motifs and Mechanisms for Oscillatory Dynamics of Biochemical Networks

Abstract:

We formulate and examine detailed mechanisms for a number of enzyme and biochemical oscillatory/signalling networks with the aim of uncovering critical subnetworks that generate the excitable/bistable/oscillatory dynamics. Such core models of the specified nonlinear dynamics are often referred to as network motifs. Our method of finding the motifs is based on a theory of stability of chemical networks known as the stoichiometric network analysis. Given that the mechanism is formulated in terms of mass action kinetics, the instabilities of steady states are linked to topological features of the (sub)network diagrams. This approach enables a classification of biochemical networks. Based on our experimental observations we examine in vitro enzyme reactions including the glucose–glucose oxidase– catalase–hydrogen peroxide reaction, the urea–urease reaction and glucose–glucose oxidase–ferricyanide reactions. Topology of the core oscillatory subnetworks are compared against one another as well as against other enzyme reactions known to oscillate, such as oxidase–peroxidase reaction. We also examine some in vivo enzyme systems of biological importance including the mitogen-activated protease kinase (MAPK) and systems determining circa- and ultradian cycles in photosythesizing microbes.

428089 Asymmetry in Signal Oscillations Contributes to Efficiency of Periodic Systems

Abstract:

Oscillations are an important feature of cell signaling that often result from complicated combinations of positive and negative feedback loops. The encoding and decoding mechanism of oscillations based on amplitude and frequency have been extensively discussed in literature in the context of inter-cellular and intra-cellular signaling [1]. To name a few, the oscillation patterns of cytosolic calcium [2], p53 tumor suppressor protein [3], NF-κB transcription factor [4] and many more have been studied in their respective networks. Although most signaling mechanisms in real physiological systems likely have an extremely complicated mode of modulation, in some cases it is possible to identify the key molecules in a given pathway to determine how one influences another. Among these cases, amplitude-to-amplitude modulation is relatively straight forward, and consequently this idea was applied early in the study of receptor-mediated signaling under sub-saturation regions [1, 3, 5]. In contrast, frequency-modulated mode of cellular signaling, especially decoding, is more convoluted. This mode of signaling has been most extensively studied in the context of calcium signaling, and while it is generally agreed that frequency is the main drive that leads to differentiated downstream events, there is also an argument that cumulative peak duration, not frequency, defines signaling specificity [6]. All of the aforementioned efforts attempt to understand in depth the oscillatory behavior in cellular and intercellular signaling. However, the fundamental question of why cellular signals appear in periodic manner as opposed to a non-oscillatory manner has not been fully answered.

In an attempt to answer this question, we designed a study to analyze oscillatory characteristics of both signaling molecule and system output. To do this, we modified two classic feedback models, the Goodwin oscillator and PER oscillator, and compared periodic signals to steady signals of identical time-averaged values. The two oscillators were selected as experimental systems because: 1) they are both self-sustained oscillators driven by negative feedback; 2) the signaling molecules of these two models exhibit opposite skewness, Goodwin oscillator having fast ascension and slow descension, and PER oscillator having slow ascension and fast descension. Furthermore, it is easy to generate a range of different intensities of skewness in both models by modifying the degree of inhibition, represented by the Hill coefficient or the cooperativity coefficient.

Currently our research into these periodic systems lends the following observations. As the inhibition of both models is increased, the resource generation develops more drastic skewness, which is correlated with an increased efficiency of the oscillatory system as compared to the output of no-feedback counterpart systems. As the cooperativity coefficient is increased in the PER model, the resource tends to skew negatively (toward the right side of the center), indicating slower ascension and faster descension of the signal. This correlates with an increased frequency and output of the system. As the Hill coefficient increases for the Goodwin model, the resource exhibits the opposite effect – it skews positively and is correlated with increased frequency and system output. Combining these results, we hypothesize the existence of regions of degree of inhibition and signaling asymmetry that will improve system output and efficiency. Furthermore, we observe a strong positive correlation between output amplitude and system output (and efficiency), and hypothesize that strong sustained oscillations give rise to a more efficient system. Our hypotheses address the fundamental question of why cellular signals and responses are periodic. From the results, we also search for the underlying mechanism that drives system efficiency.

[1] M. Behar and A. Hoffmann, "Understanding the temporal codes of intra-cellular signals.," *Curr. Opin. Genet. Dev.*, vol. 20, no. 6, pp. 684–93, Dec. 2010.

[2] E. Smedler and P. Uhlén, "Frequency decoding of calcium oscillations," *Biochim. Biophys. Acta - Gen. Subj.*, vol. 1840, no. 3, pp. 964–969, 2014.

[3] M. Lev Bar-Or, R., Maya, R., Segel, L.A., Alon, U., Levine, A.J. and Oren, "Generation of oscillations by the p53-Mdm2 feedback loop: A theoretical and experimental study," *Proc. Natl. Acad. Sci.*, vol. 97, pp. 11250–11255, 2000.

[4] R. Williams, J. Timmis, and E. Qwarnstrom, "Computational Models of the NF-KB Signalling Pathway," *Computation*, vol. 2, no. 4, pp. 131–158, 2014.

[5] J.W. Black, "Operational models of pharmacological agonism," *Proc R Soc L. B Biol Sci*, vol. 220, pp. 141–162, 1983.

[6] L. Zhu, S. Song, Y. Pi, Y. Yu, W. She, H. Ye, Y. Su, and Q. Hu, "Cumulated $Ca2^{+}$ spike duration underlies $Ca2^{+}$ oscillation frequency-regulated NFκB transcriptional activity.," *J. Cell Sci.*, vol. 124, no. Pt 15, pp. 2591–601, Aug. 2011.

430552 Dynamic Modeling of Molecular Motor Association with Axonal Amyloid Precursor Protein Vesicles

Abstract:

Mano Ram Maurya[1], Lukasz Szpankowski[1,2], Lawrence S.B. Goldstein[3], Shankar Subramaniam[4]

ABSTRACT

Targeted transport of vesicles, organelles, and other types of intracellular cargo along microtubule tracks is powered by kinesin and cytoplasmic dynein molecular motors. Both classes of motors can attach to the same cargo and thus their binding kinetics decides the distribution of the cargo with these motors. In particular, we have been studying the binding of these motors to amyloid precursor protein (APP) vesicles in mouse hippocampal axons. It has been shown experimentally that several units of each of kinesin and dynein proteins can be bound to APP vesicles. In the present work, we have developed a 16-state biochemical reaction network-based model to capture the dynamics of binding of kinesin and dynein to APP vesicles. Each state represents an APP vesicle with a specified number of associated kinesin and dynein motors, up to a maximum of three units for each. A set of nonlinear coupled ordinary differential equations were generated describing the dynamic mass-balance based on the reaction rates for association/disassociation of each motor species. The system of differential equations was solved to compute the steady-state contribution of each axonal APP vesicular cargo associated with a defined arrangement of molecular motors. Since the rate parameters were not known *a priori*, the endogenous association/dissociation rate of both kinesin and dynein on APP vesicles was estimated through a nonlinear optimization (data-fitting) approach. The resulting model predicted the dynamics in under- and over-expression studies, has been validated against experimental data in heterozygous knockout kinesin mutants, and facilitates prediction of transport dynamics in systems not amenable to experimental manipulation. Furthermore, the systems biology approach used here provides a framework to build detailed predictive kinetic models to characterize the molecular states using macro-level data, which can then be used to generate and test hypotheses leading to the design of novel experiments and further refinement of such models.

Acknowledgements: We would like to acknowledge the National Science Foundation (NSF) collaborative grant STC-0939370.

Key words: mechanistic dynamic modeling, parameter estimation, systems biology, vesicle transport, kinesin, dynein.

---

1 Equal effort.

2 Current address: Fluidigm Corporation, 7000 Shoreline Court, Suite 100, South San Francisco, CA 94080.

3 Corresponding author. E-mail: lgoldstein@ucsd.edu, Phone: (858) 534-9700, Fax: (858) 246-0162.

4 Corresponding author. E-mail: shankar@ucsd.edu, Phone: (858) 822-0986, Fax: (858) 822-3752.

432303 Phenotypic and Gene Expression Variability Underlies Adaptive Resistance in Heterogeneous Bacterial Populations

Abstract:

The rapid rise of drug-resistant superbugs and the declining antibiotic pipeline are serious challenges to global health. The root cause of the antibiotic resistance crisis is the ability of bacteria to evolve tolerance to antibiotics and other environmental toxins. Bacteria exhibit extensive intra- and inter- population heterogeneity, which provides an advantage in adaptation to diverse environmental pressures, but renders identification of underlying mechanisms of adaption challenging. Here, we used RNA-Sequencing to locate key regulators of adaptive resistance by investigating transcriptomes of bacterial populations upon prolonged exposure to diverse chemical toxins and comparing to gene expression profiles in unadapted populations. Our lab-evolution experiments show that initially antibiotic-sensitive isogenic bacteria adapted to the same toxin exhibit rapid population-level phenotypic divergence. To investigate the mechanisms underlying heterogeneity across populations, we evolved clinically derived multi-drug pathogens and lab strains of *Escherichia coli* to a range of dissimilar toxins, including antibiotics and biofuels. We find a set of genes that are differentially expressed across multiple adaptation conditions. Further, we observe the presence of significant shifts in gene expression variability across unadapted and adapted populations. A more focused quantification of mRNA expression of stress-response genes corresponding to the multiple antibiotic resistance (*mar*) regulon, the general stress response, and the SOS response, all known to regulate adaptation, confirms the underlying gene expression heterogeneity. Hierarchical clustering and principal component analysis reveal that adapted populations exhibit variability in transcript levels that is distinct from unadapted populations; certain stress response genes are observed to increase or decrease expression variability with respect to wild type, suggesting the presence of dissimilar regulation of gene expression during adaptation. By propagating strains with corresponding stress response genes either knocked out or perturbed via CRISPR-interference mechanism, we identify unique relationships between gene expression variability and impact on adaptation that differentiate adapted or unadapted populations. This study provides insight into the dynamic nature of the bacterial adaptive transcriptome and evidence that gene expression variability can be used to identify bacterial populations undergoing adaptive resistance. The identification of genes contributing to adaptive resistance may be an avenue towards the development of antimicrobial agents that hinder the onset of drug-resistance.

432675 Understanding the Poplar Microbiome Structure in Response to Host Stress

Abstract:

Plant roots in the soil provide a rich carbon source through root exudates and organ turnover that attracts soil bacteria. The bacteria associated with plants are diverse, with thousands of unique bacteria measured as operational taxonomic units (OTUs) identified in natural samples. The associated bacteria can have beneficial effects on the host plant through nutrient acquisition, signaling and defense, and buffering of extreme environmental conditions. The plant microbiome has been shown to be dependent on host plant genotype, soil type, and environmental conditions. Given the importance of plants as feedstocks for food and energy, it is important to understand how the microbiome changes in response to natural stresses and how these changes may supplement the host response.

Using *Populus* trees as host plants, we investigated how abiotic stresses affect the host plant and belowground microbiome. Drought stress was implemented by watering plants at complete wilting to identify potential microbes that are beneficial in buffering the plant against drought. Drought plants were significantly reduced in overall growth, with extreme decreases in photosynthesis rate due to closed stomata especially at the lowest water conditions. The effects of drought on the host plant and environment significantly changed the endophytic microbiome relative to control, with increases in alpha and gamma proteobacteria. Only 15 of the top 20 OTUs were shared between control and drought plants, representing 73% and 65% of the total identified OTUs in the system, suggesting that drought microbiome is more diverse relative to controls. In parallel, a shade stress was induced by limiting light input to near the CO_2 compensation point, affecting carbon allocation and the chemical environment of the host, ultimately leading to changes in the microbiome. 16 OTUs were shared with the control top 20, representing 65% of the total diversity. 4 of the differential OTUs were different than what was observed in drought stress. The microbiome of shade and drought plants clustered together by principal coordinate analysis, suggesting a core microbiome that is enriched under host stress conditions. With this information we can begin to identify bacteria that are conserved in all *Populus* host growth conditions, as well as bacteria that enriched under specific environmental stresses. This research will lead to holistic understand of how the microbiome community functions to supplement the host phenotype.

430660 Understanding Candida Albicans and Candida Glabrata Mixed Species Biofilm Growth and Co-Culture

Abstract:

Candidiasis, fungal infection caused by *Candida* species, affects patients of all ages and ranges from thrush in the oral cavity to disseminated deep tissue infections. *Candida* infections may be difficult to treat due to inherent or acquired antifungal resistance, and limited number of available therapeutic agents; with immunocompromised patients being the largest at-risk population. Biofilms, which are often associated with the pathogenesis of *Candida*, are generally polymicrobial in nature, and are held together by an extracellular matrix (ECM) that is attached to a surface. In the human host, this surface may by a catheter, mucosal surface of the gastrointestinal tract, or lining of the stomach. Biofilms are difficult to treat largely due to the protective ECM, which reduces the penetration and efficacy of antifungal agents. *C. albicans* and *C. glabrata* are the two most commonly isolated species from *Candida* infections, and have been isolated together, which makes polymicrobial populations an interesting and clinically relevant area of study. In this work, we characterize the impacts of co-cultures in biofilm formation and population dynamics using mixed cultures of *C. albicans* and *C. glabrata*. When grown as biofilms, we find that biofilm formation of the polymicrobial population depends on the relative ratio of starting cell concentrations of *C. albicans* and *C. glabrata* in an unexpected manner. The specific ratio we find to yield the highest biomass is when the ratio of *C. albicans* to *C. glabrata* is low. *C. glabrata*, in comparison to *C. albicans*, is not very successful at forming biofilms when grown as a single species. Interestingly, we find that the higher ratios of *C. albicans* to *C. glabrata* in fact result in less biofilm formation than the low *C. albicans* to *C. glabrata* counterpart. When co-cultured in liquid medium in suspension, one of the species consistently exhibits a growth advantage over the other. Our results suggest interspecific interactions between these two fungal pathogens, where a polymicrobial population causes unexpected changes in biofilm formation. The underlying mechanisms of these inter-specific interactions are being investigated.

437407 Engineering Multicellular Motility-Based Biosensors Systems

Abstract:

The ability to robustly control cellular motility in response to an external stimulus can facilitate the development of biosensors or the construction of micron-sized robots that use microbes to enable movement. By separating the sensing and actuating cells, and coordinating their behavior via cell-cell communication, we can improve system design through modularity, signal amplification and logic gate-type responses to multiple inputs. Our first goal was to use a synthetic biology approach to build a system where the motility of *Escherichia coli* is controlled by a cell-cell communication signal. Towards this goal, we have designed and constructed a system where *E. coli* motility is controlled by an acyl homoserine lactone (AHL) quorum sensing (QS) signal molecule. Unlike previous efforts that used transcriptional activation-based QS systems to control motility, we used the repressor-based esa QS system to enable tight regulation of a key motility gene. We have demonstrated that we can restore motility in an *E. coli* strain with a *motA* deletion by expressing *motA* under the control of the esa promoter, which is tightly repressed by the QS transcriptional regulator (EsaR) in the absence of AHL. Expression of *motA* is triggered by AHL, which caused EsaR to dissociate from the *esa* promoter. To quantify motility, we measured the swimming of cells through semi-solid agar plates. We observed that cells do not swim in the absence of AHL, and swimming is induced in the presence of micromolar concentrations of AHL. Overall, our engineered communication-dependent motility (CoMot) cells show AHL dependent motility. We replaced wild type EsaR with a mutant that responds to nanomolar AHL concentrations to increase the AHL sensitivity of the cells (CoMot+). Interestingly, directional swimming of the CoMot and CoMot+ cells was observed up an AHL gradient even though chemotaxis was not directly controlled. Further, we have demonstrated that the cells respond to AHL produced by a second strain of *E. coli.* When the AHL production is inducible, the combination of AHL-producers and the CoMot cells enable a novel biosensing framework. The CoMot strains engineered here can be used to build synthetic networks and biosensor systems where a motility output is tightly regulated and controlled by cell-cell communication.

429325 Improving the Residence Time of Probiotics in the Gut through Functional Metagenomics

Abstract:

The human large intestine houses trillions of microorganisms which collectively form the highly diverse microbial community known as the gut microbiota. The gut microbiota is becoming increasingly recognized as an organ, performing many functions critical to the maintenance of health, including extraction of nutrients, production of vitamins, and defense against pathogens. Like an organ, pathologies in the structure of the gut microbiota (termed "dysbioses") can have significant negative impacts to the human host, including obesity, malnutrition, cancer, and death. However, the gut microbiota is currently unique among the organs in that it is highly engineerable, enabling improvement of function through the substitution of beneficial microbes for less desirable ones. Towards this goal, probiotics are orally ingested microorganisms which improve gut health through provision of additional metabolic functions or through competition for niche space with pathogens. Unfortunately, current probiotic strains have short residence times in the gut, necessitating continual administration for treatment of chronic conditions. In order to improve the gut residence time of the important probiotic *E. coli* Nissle, we extracted tens of gigabases of metagenomic DNA from several highly distinct gut microbial communities and selected colonization-enhancing gene fragments via passage in both gnotobiotic and conventional mice. Bioinformatic and biochemical analyses of these fragments indicated mechanisms of improved colonization, and integration of these colonization factors into probiotic *E. coli* Nissle significantly improved overall gut residence time. Subsequently, these fragments were tested in combination to investigate synergistic and antagonistic effects between colonization factors. Taken together, this work has the potential to significantly improve our understanding and use of probiotics while laying an important foundation for engineering the human gut microbiota.

420101 Agent-Based Models of Heterogeneous Tumor Populations Predict Multi-Scale Dynamics and Inform Therapeutic Perturbations

Abstract:

Cancer accounted for 8.2 million deaths in 2012, a number that continues to grow despite the efforts of researchers and medical professionals alike (World Health Organization, 2014). The challenge in understanding cancer, and thereby developing successful therapies, stems from its inherent heterogeneity and the complexity of its microenvironment. It is therefore essential to study, and account for, the manner in which tumors interact with the environment. To address this need, we developed a tunable, multi-scale *in silico* model that predicts the emergent behavior of a tumor in response to a variety of environmental and therapeutic perturbations.

Continuous and deterministic models cannot account for the discrete, stochastic, and heterogeneous nature of tumor cells, prompting researchers to consider a different paradigm: bottom-up rather than top-down modeling. Bottom-up approaches characterize the low-level building blocks of a system and couple them together to predict dynamic responses. Here, these building blocks, or agents, are individual healthy or cancerous cells. These cells are coupled in a 3D microenvironment that accounts for crowding, nutrient resources, and cytokine signaling (both secretion and uptake). Unlike cellular automata (CA) models, the agent-based modeling (ABM) framework allows both spatial and temporal resolution so these cells can divide, invade, and apoptose across space and time.

Our multi-scale ABM incorporates inter- and intra-cellular interactions, nutrient diffusion, and paracrine signaling. Cell agents, both healthy and cancerous, are equipped with a simplified, 14-component EGFR network that interacts with the environment (Zhang et al., 2009). Each cell agent can exist in one of five phenotypes—quiescent, proliferative, invading, apoptotic, or necrotic—and follows rules defined by its phenotype, local environment, and neighboring cells. More specifically, the intracellular network is most responsible for proliferation and invasion, and—through the corresponding dynamics—couples the cell to the physical environment. Glucose availability and crowding is most responsible for a cell's entrance to, and exit from, quiescence. Cancerous agents have a modified rule set based on the "hallmarks of cancer" described by Hanahan & Weinberg (2000). Examples of distinctions include increased tolerance for crowding and an accelerated cell cycle.

The flexibility of agent-based modeling is demonstrated by the ease with which new agents and perturbations can be introduced. A sensitivity analysis of the model demonstrated significant sensitivity of parameters affecting glucose conditions, highlighting glucose concentration as a potential control input. Variable glucose conditions, in combination with existing treatment strategies—excision, radiation, and chemotherapy—will be simulated to predict therapeutic efficacy. These tunable perturbations can be introduced at any time point and in any combination. Preliminary results quantify the effects of increasing excision radius under different glucose conditions, revealing important tradeoffs that should be explored in future experiments as well as strategies to be validated in the clinic. The development and analysis of tumor growth under different parameters and therapies demonstrate the versatility of ABMs for guiding translational cancer therapy and informing more personal medicine.

Hanahan, D., & Weinberg, R. A. (2000). The hallmarks of cancer. *Cell*, **100**(1), 57-70.

World Health Organization. Global status report on noncommunicable diseases. 2014.

Zhang, L., Want, Z., Sagotsky, J. A., & Deisboeck, T. S. (2009). Multiscale agent-based cancer modeling. *J Math Biol*, **58**, 545-559.

433845 Mathematical Modeling of Podocytes in Diabetic Kidney Disease

Abstract:

Diabetic kidney disease (DKD) is the primary cause for end-stage renal failure, is among the severe complications of diabetes, and is a significant factor leading to morbidity and mortality in both type I and type II diabetic patients. Hyperglycemia is known to initiate and exacerbate the pathophysiology of DKD. In the early stages of DKD, hyperglycemia induces increases in the rate of blood filtration through capillary beds (glomeruli) within the filtration units (nephrons) of the kidney [1]. The increase in glomerular filtration rate (GFR) contributes to glomerular injury, eventual damage to the tubular sections of nephrons downstream of the glomeruli, and loss of renal function. Damage to the epithelial cells called podocytes surrounding the glomerular capillaries is a critical factor in glomerular injury but is not immediately detectable with non-invasive clinical measurements of urine albumin concentration until after proteinuria develops due to substantial leakage of proteins through damaged glomeruli [1 – 2]. However, it is important to slow the rate of progression before irreversible loss of podocytes and nephrons occurs. In this work, we use mathematical modeling to investigate factors that contribute to the onset of nephropathy in DKD and seek to predict nephropathy in advance of detection of proteinuria by describing and simulating the mechanism of podocyte damage and loss. While mathematical models have been proposed for various renal physiological and pathophysiological processes, a quantitative description of the dynamics of podocyte injury at the onset and during the progression of DKD is lacking in the published literature. We have formulated a mathematical description of the mechanism of podocyte injury in DKD by including the contributions from glucose metabolism in the kidney during hyperglycemia, nephron hemodynamics, podocyte cell-signaling, and the dynamics of biomechanical forces that actively and passively alleviate or exacerbate effects of damage. These processes interact across multiple length and time scales. The mathematical model consists of a system of ordinary and partial differential equations and algebraic equations that are solved numerically. The intensity and duration of elevated GFR during periodic hyperglycemia is varied in simulations to investigate the influence on the transition to irreversible podocyte loss. The simulation results from the mathematical model are compared to published experimental data to assess the model ability to accurately and quantitatively predict how podocyte loss depends on filtration rate and the onset of proteinuria after hyperglycemic events.

[1] Lopez-Novoa JM, Martinez-Salgado C, Rodriguez-Pena AB, Lopez-Hernandez FJ. Common Pathophysiological Mechanisms of Chronic Kidney Disease: Therapeutic Perspectives. Pharmacol Ther 2010;128:61-81.

[2] Garg P, Holzman LB. Podocytes: Gaining a Foothold. Exp Cell Res 2012;318:955-63.

430845 Genome-Scale Modeling of the Core Proteome of Metabolism and Expression, and Its Transcriptional Regulation

Abstract:

The cell 'chassis' is of fundamental importance for synthetic biology applications. To predictively engineer this chassis, a systems biology understanding of it is critical. In this work, we broaden our understanding of the chassis using integrated genome-scale models of metabolism and expression. We first define a core proteome consisting of 356 gene products, accounting for 44% of the *Escherichia coli* proteome by mass based on proteomics data. This core proteome includes 212 genes not found in previous comparative genomics-based core proteome definitions, accounts for 65% of known essential genes in *E. coli*, and has functional overlap with minimal genomes (*Buchnera aphidicola* and *Mycoplasma genitalium*) (AUC = 0.78). Based on transcriptomics data across growth conditions and genetic backgrounds, the systems biology core proteome is significantly enriched in non-differentially expressed genes, and depleted in differentially expressed genes. Compared to the non-core, core gene expression levels are also similar across genetic backgrounds. Furthermore, core genes exhibit significantly more complex transcriptional and post-transcriptional regulatory features (40% more transcription start sites per gene, >20% longer 5'UTR). Thus, genome-scale systems biology approaches rigorously identify a functional core proteome needed to support growth. This framework, discerned and validated using high-throughput datasets, facilitates a mechanistic understanding of systems-level core proteome function through in silico models. We then investigate how the core proteome is regulated using a genome-scale, transcriptional regulatory network of *E. coli*. Taken together, the tightly regulated core proteome of *E. coli* comprises a fundamental repertoire of genes and biological functions that are of importance to a broad range of microbial synthetic biology applications.

423632 Investigation of the Antibiotics-Triggered Perturbation in the Metabolism of Pseudomonas Aeruginosa Via a Metabolic Modeling Approach

Abstract:

Although antibiotics are known to eliminate pathogens through a variety of mechanisms [1, 2], it is reported that antibiotics may induce the transition in the microbial phenotypes and promote the formation of persisters, which are dormant variants of regular cells that are highly tolerant to antibiotics [3]. For example, Dr. Wood's group at Pennsylvania State University reported that *E. coli* pretreated with antibiotics such as rifampicin, tetracycline, and carbonyl cyanide m-chlorophenyl hydrazine increased the microbial persistence dramatically. They demonstrated that the bacterial persistence resulted from halted protein synthesis and a variety of environmental cues [4]. The presence of persister-specific resistance is suggested to account for the recalcitrance of infectious diseases [5, 6]. It is thus important to study the mechanisms via which antibiotics cause pathogens to change their metabolism to form persister cells. Experimental approaches such as transposon insertion have been used to study the phenotype transition mechanisms during antibiotic treatment. Screening knockout libraries has not produced mutants that lack persisters, indicating that dormancy mechanisms are not regulated by single genes or enzymes [3, 7]. This understanding motivated us to develop a systems level approach that can incorporate the interaction of multiple genes/enzymes/metabolites to investigate potential persister-forming mechanisms triggered by antibiotics. Since genome-scale models contain the genes/enzymes/metabolites for metabolic pathways that determine the microbial growth rate, they can be used to pinpoint the genes/metabolites involved in the phenotypic transition. In particular, some metabolites in genome-scale models behave like signaling molecules in triggering persister formation. One example is acyl-homoserine lactone, which promotes the persister formation in *Pseudomonas aeruginosa* through quorum sensing [8]. Studying the metabolic alternation between different phenotypes may provide cues to potential signaling metabolites and pathways to manipulate the formation of persister cells. Although research has been conducted on the reaction-flux distribution upon the antibiotic treatment [9], no genome-scale modeling approach has been published to quantify the effect of antibiotics on bacterial phenotype transition.

In this work, we present the first genome-scale modeling approach to quantify the effect of antibiotics on the metabolic flux redistribution, and thus investigate potential mechanisms for the persister formation that is triggered by antibiotics. Specifically, we first developed an approach to integrate gene expression data with metabolic models to identify the reactions that positively reflect the metabolic activities of microorganisms. On the basis of these metabolic-activity indicating reactions, we then developed the first genome-scale approach to predict microbial metabolic perturbation of *P. aeruginosa* upon antibiotic treatment. *P. aeruginosa* is selected as the example microorganism in this work, as it is one of the leading causes of nosocomial infections in hospitalized patients and it displays resistance to a wide array of antibiotics by forming a biofilm in chronic infectious processes [10, 11]. The results showed that 1110 metabolic reactions within the network, 171 reactions increased flux, 372 reactions decreased fluxes and 10 reversible reactions changed flux direction. Parts of our results were further validated by experiment data related to persister cell formation. The flux profile of cell with antibiotic treatment provides clues on the detection of novel signaling pathways that trigger cell persistence. The exchange rates of several metabolites with significant alternation were also proposed. Specifically, 14 metabolites including acetate, fumarate, gluconate, succinate, citrate, ethanol, glycerol, and pyruvate increased their exchange rates, while 6 metabolites including 4-hydroxybenzoate, orotate and allantoin decreased their exchange rates. These metabolites may be used in further controlling persister formation in nutrition supplement. Given the metabolic network information, this approach can be applied to study the metabolism adjustment in phenotype transition of other microorganisms.

433487 A Modular Computational Framework for Multi-Scale Simulation of Chemotaxis in Complex Environments

Abstract:

Chemotaxis has long served as a model system for signal transduction in both bacteria and eukaryotes. Numerous mathematical models have been developed over the years to explore different aspect of these pathways. These models have tended to focus either on the intracellular mechanisms governing pathway dynamics or the coarse-grained, population-level behavior of swimming/migrating cells in chemical gradients. Only a few models have attempted to link the two. In this work, we developed a modular multi-scale computational framework for simulating cell behavior in complex chemical environments using a hybrid-continuum approach. This framework enables us to directly employ intracellular pathway models for individual cells to determine population-level, macroscale cell behavior. Our strategy is based on the particle-in-cell approach. Here, the aggregate effect of many cells in a Lagrangian frame of reference is interpolated onto Eulerian (stationary) mesh points, where the moments of the continuum distribution (e.g. concentration) are also computed and updated. The typical cycle involves the following: (i) integration of the pathway equations of motion for individual cells; (ii) interpolation of the cell effects (e.g. metabolism) onto the stationary mesh; (iii) computation of the new field values using this information; and (iv) interpolation of the fields from the mesh back to the particle locations. To validate this framework, we investigated bacterial chemotaxis in complex chemical and spatial environments. As we demonstrate, our simulation results are quantitatively in agreement with the corresponding experimental results. While this framework was developed to study chemotaxis, we will also illustrate how it can be used to simulate at multiple scales any biological system involving cell populations interacting via chemical signaling.

434178 Spatial Patterns of Growth and Inhibition in Spreading Infections of Vesicular Stomatitis Virus

Abstract:

Viruses infect their multi-cellular hosts through processes that play out over multiple scales: a virus particle infects a susceptible host cell, producing virus progeny that then spread to other cells within the host and initiate new rounds of infection. Infected cells make not only virus progeny, but also virus-like defective interfering particles (DIPs), which lack essential genes needed for replication. DIPs are found in natural isolates of influenza A, dengue and West Nile viruses, but their roles in infection spread and disease pathogenesis are not known. DIPs alone are non-infectious. However, in co-infections with intact virus, DIP genomes can interfere with virus growth by competing for viral polymerase, replicating at the expense of viral genomes, or by binding packaging proteins, assembling DIP progeny at the expense of intact virus particles. To elucidate how DIP-virus interactions play out in spreading infections we engineered wild-type and DIP strains of vesicular stomatitis virus to encode and express red and green fluorescent proteins, respectively. Moreover, the reporter-DIP was further designed to carry a deletion of the gene for the viral surface glycoprotein G (VSV-ΔG-eGFP). With this system we were able to monitor the dynamics of wild-type and defective virus gene expression through their respective reporter proteins in single co-infected cells. In co-infections of wild-type (RFP) the reporter-DIP inhibited production of infectious virus to the same extent as standard non-reporter DIPs, but reporter-DIPs were less potent in their ability to inhibit wild-type (RFP) reporter protein production. Heterogeneity in protein expression patterns was apparent among cells co-infected with the same dose of wild-type and reporter-DIPs, but wild-type protein expression (RFP) overall decreased as DIP reporter protein (GFP) expression increased. When infections were permitted to spread spatially across a population of susceptible cells, the heterogeneity in reporter protein increased. Unlike spreading infections in the presence of natural DIPs, which can completely arrest infection spread, co-infection with reporter-DIP produced many local areas of wild-type virus escape, and complete arrest of spread was not observed. These results suggest that competition for packaging protein G can reduce wild-type virus production at the cell level to the same extent as natural DIPs, but this reduction is not sufficient to fully inhibit wild-type virus spread in host cell populations. More broadly, these results highlight how similar patterns of virus growth and inhibition at the level of individual cells can exhibit distinctly different properties of infection spread or containment when virus and DIPs propagating spatially across a population of susceptible cells.

432782 In silico Dynamic Gene Expression Models Quantify the Impact of Regulatory Logic Gates on Network Inference

Abstract:

Network inference algorithms are widely used to infer biological networks, but their capacity to resolve regulatory logic gate relationships remains largely unknown. In this study, we built an *in silico* model of dynamic gene expression to quantify the influence of logic gate regulation on the accuracy and confidence of network inference algorithms in order to elucidate the strengths and limitations of inferring gene regulatory networks.

Network graphs are a simple and intuitive means to represent the underlying topology and information transfer inherent to gene regulatory systems. Accurate network models provide valuable insight into functional modules and information flow, highlighting targets for therapeutic intervention. With high throughput technologies, network inference has become a conventional method to uncover regulation among genes and identify strategies to modulate system behavior. However, every method has its limitations: the DREAM consortium recently demonstrated specific network motifs, and method of inference, impact accuracy of the resulting inferred network model[1]. A significant source of this error is likely attributed to regulatory logic gates and its associated kinetic parameters, neither of which are considered in common practice.

An *in silico* model was built to simulate a gene regulatory network. Six different networks–each containing five nodes —define the activation and inhibition among genes. Genes with multiple inputs are assigned an AND, OR, NAND, or NOR logic gate relationship. A thermodynamic ODE model generates dynamic gene expression data based on the network interactions and logic gate relationships. GENIE3, a well-established inference algorithm that can deal with combinatorial and non-linear interactions, was selected to infer the putative gene regulatory network[2]. Each *in silico* data set was inferred 100 times using GENIE3 and scored against a permuted data set in order to calculate each edge's statistical significance. The network inference and scoring objective were performed over a range of 441 logic gate kinetic parameters

The GENIE3 algorithm infers AND-gates with a higher confidence than OR-gates for 94% of the tested kinetic parameter space for the bi-fan network structure. In the feed-forward network structure, the AND-gate relationship is inferred with a higher confidence for 90% of the kinetic parameter space. In the bi-fan network, GENIE3 infers OR-gates with higher confidence when both inputs into a single gene have low kinetic parameters. However, for the feed-forward loop, the OR-gate is inferred with a higher confidence when only one edge within the network motif has a low kinetic parameter value. Lastly, GENIE3 shows little inference bias between NAND and NOR gates. Further simulations will evaluate Bayesian, regression, and correlation based inference algorithms, as well as the SUM logic gate relationship.

GENIE3, one of the top performing inference algorithms in the DREAM challenge, demonstrates bias when inferring specific logic gate relationships, limiting its predictive insight. For instance, in drug design, it is critical to consider biases in an inferred mechanism because AND-gates would be targeted and perturbed differently than OR-gates. These biases would also be critical in designing combinatorial drug targets and predicting the impact of upstream perturbations on the mechanism of interest. By considering the limitations and biases of network inference algorithms, we will be able to build more informed network models for the future design of drug targets or perturbation studies from inferred network models.

References:

(1). Marbach, Daniel *et al*. "Wisdom of Crowds for Robust Gene Network Inference." Nature Methods 9,8 (2012): 796-804

(2). Huynh-thu, Van Anh *et al*. "Inferring Regulatory Networks from Expression Data Using Tree-Based Methods." PLoS ONE, 5(9): e12776, 2010.

432630 Coarse-Graining of Heterogeneous Neural Dynamics

Abstract:

In biology, the structure of networks plays an important role in the resulting dynamics both intracellularly and in the interactions between cells, especially in the nervous system. The realism of simulations of large networks of neurons can be enhanced by introducing cell heterogeneity: neurons are characterized by different values of physiological parameters, as well as by different connectivities in the network. In this case, we find that the (evolving) system state often quickly becomes a smooth function of these parameters. We make use of this phenomenon to express the system state through a significantly reduced number of variables, allowing for acceleration of several computational tasks.

The problem of coarse-graining a large heterogeneous network of neurons thus becomes that of finding the appropriate basis functions (the appropriate "observables") given the (multidimensional) distribution of the parameters. We use nonparameteric techniques to generate such basis function sets which are orthogonal with respect to arbitrary multidimensional parameter spaces. For high-dimensional spaces of heterogeneous parameters, we use analysis of variance (ANOVA) as an alternative coarsening technique.

With a coarse description, we show how to accelerate tasks such as finding coherent steady states and synchronized limit cycles. We demonstrate applications of these techniques both for networks of simple coupled oscillators, and for a much more complicated simulation of circadian dynamics in the suprachiasmatic nucleus of the mammal hypothalamus.

415023 Effects of Nonlinear Responses of the Huang-Ferrell Model of MAPK Cascades on the Character of Autocrine Communication

Abstract:

Autocrine communication in cellular systems is mediated by the interaction between extracellular ligands and membrane-bound receptors. In our previous works, we have studied the effects of ligand-receptor kinetics, transport parameters, and interstitial flow on the propagation velocity of autocrine signaling. However, the process of signal transmission strongly depends on the dynamics of intracellular processes. Signal coming from a ligand-receptor complex into the intracellular space is typically amplified in MAPK phosphorylation/dephosphorylation cascade. It was recently shown that the widely accepted Huang-Ferrell model of MAPK signaling offers bistable (switch-like) and oscillatory behavior. These nonlinearities can strongly alter the release of new ligand molecules into intracellular matrix as the MAPK output modifies various transcription factors or activates protease precursors. Here we developed a mathematical model of autocrine communication that considers important reaction-transport processes in the intracellular and extracellular domains together. By the method of stoichiometric network analysis, we have found the minimal subnetwork of MAPK cascade that offers oscillatory behavior and steady state multiplicities. This subnetwork was then incorporated into the model of autocrine signaling, where it plays a role of positive or negative feedback for the process of ligand release. By numerical simulation of the model equation, we study transient behavior of the developed model at the interface between cells with and without high MAPK activity. We identify and classify possible responses of cellular systems to specific regimes of MAPK cascades in parametric space.

422070 Transient Dynamics of a Four State Actomyosin Model of Muscle Contraction

Abstract:

A model of muscle action has been used to compute the steady state response of mechanical and chemical variables in muscle action. the merits of a modelof this action, however, depends on both itssteady state and transient resonses. This model depends on a posted high strain and low strain state for the actomyosin *ADP complex (AMADP). Combining these two strain states with a simple averaging rule and mass action kinetic models for the biochemical cycleallows for computation of both steady state and trasient responses.The overall theory and the steady state responses of this model were published (Long and Phoonsiri, 2006). The transient responses are considered here.Compliance corrections have not yet been included in the model.The model simply links the nono-scale to the macro scale using the half sarcomere as the basic system for analysis.

A comparison of computed results is made with well-known mechanical tests widely published in the literature. In addition to mechanical tests, a number of species transients are also presented. These pose additonal possible results that could be compared against experiment once the experiments are done.

409093 An Engineering Approach to Discover Novel Antivirulence Strategies

Abstract:

Antibiotic resistance is a growing public health threat that is worsened by a declining antibiotic pipeline. Strains resistant to "last line of defense" antibiotics have begun to appear and the danger of our antibiotic arsenal becoming obsolete is quite real. Since current antibiotic discovery has failed to keep pace with the rise of resistance, novel methodologies are required to address this looming crisis. Antivirulence therapies have been receiving increased attention as a novel class of anti-infectives. Rather than targeting essential bacterial functions, as current antibiotics do, antivirulence therapies target essential host-pathogen interactions required for infection such as adhesion, quorum sensing, and susceptibility to immune attack. These therapies are less prone to resistance development than are antibiotics, and have the potential to fill the widening therapeutic gap caused by the rise of multi-drug antibiotic resistance.

Reactive oxygen and nitrogen species (RNS, ROS) are antimicrobials generated by immune cells to kill bacteria. The importance of RNS and ROS to immune function is evidenced by the many pathogens, including *Mycobacterium tuberculosis*, *Neisseria meningitides*, *Vibrio cholerae*, *Salmonella enterica*, and enterohemorrhagic *Escherichia coli* (EHEC), that depend on ROS and/or RNS detoxification to establish or sustain an infection. Inhibitors of these defense systems promise to be potent antivirulence therapies; however, known agents are either toxic to humans or poorly transported into bacterial cells. In this talk, I will discuss how we use methodologies from systems biology and metabolic engineering to quantitatively study RNS and ROS stress in bacteria, as well as how we have used the resulting knowledge to identify novel ways to sabotage the RNS and ROS defense systems bacteria require for infection.

432001 Communication and Collaboration in Synthetic Microbial Communities

Abstract:

The biosynthesis of compounds of medical and industrial importance often requires engineering and optimization of complex metabolic pathways. Traditionally, these processes have employed a clonal population of recombinant microbes. There are many limitations of using a single population of microorganism that could be alleviated or addressed by using a mixed community, including metabolic load and the number of exogenous elements that can be cloned and optimized in a single cell. In order to control interactions required for these cells to work together, it is important to generate robust communication pathways between biotechnologically relevant species. To this end, we have developed new regulatory components and systems that can be used to turn gene expression on or off in response to a cell-cell communication signal, and enable interspecies communication between representative Gram-negative (*Escherichia coli*) and Gram-positive (*Bacillus megaterium*) organisms. We are also employing new strategies to control the growth and coexistence of species in co-cultures by manipulating ecological interactions; a key challenge for applications with microbial consortia. Finally, I will describe our recent efforts to distribute functions between these microbes in a model bioprocess. This work has yielded a set of key technologies enabling the division of labor between specialized species and strains for bioprocessing and a broad range of other applications.

409204 Systematic Mining of Noncoding RNAs for Regulating Complex Phenotypes

Abstract:

An important question in strain engineering is: *how to search the genome for targets that regulate a phenotype of interest?* Given the natural role that noncoding RNAs play in coordinating complex stress-responses, there is increasing interest in exploiting these regulators to construct phenotypes of interest that can further expand the current potential for microbial cells. An emerging need in this frontier is the proper identification and selection of functional RNAs that can be engineered for optimal regulation of desired genes and relevant metabolic pathways. Thus far, the number of mechanisms and RNA scaffolds that are being exploited for these applications are highly constrained. Although a large number of "omics" approaches are being attempted to expand the number of RNAs that globally regulate phenotypes, it is unclear how to interpret this *massive* amount of data in the context of biological pathways and utilize it to uncover underlying biological mechanisms that can be relevant to strain engineering.

This talk will address characterization tools that we are developing to automate the identification of global regulators that could be exploited in constructing synthetic schemes to regulate complex phenotypes. Specifically, we will present a general pipeline for mechanistically understanding and engineering desirable complex phenotypes based on a *rational and predictive* systems analysis of global regulation and on the intracellular assaying of relevant targets. Our approach has been applied to microbial organisms with extremophilic traits and has combined bioinformatics, genetics, biophysical models and synthetic parts for prediction of pathways.

427263 The Vision of a Sustainable Bioeconomy in an Era of $50 Oil

Abstract:

Biotechnology is rapidly coming of age as enabling technology for the production of biobased chemicals and biofuels. This is a process of creative destruction whereby one type of manufacturing technology based on chemistry is gradually being replaced by another based on biology. Sustainability is a major driving force, but, biobased processes have additional unique advantages, such as superb specificity and ability to convert with high efficiency renewable feed stocks. These advantages tend to counterbalance the high space times of competing chemical processes. A critical question is how biotechnology will fare in a period of declining energy and fossil feedstock prices. I argue that, despite temporary excursions, the fundamental relationship between real costs of fossil and renewable feedstocks will be maintained and the outcome of the competition between biotechnology and chemistry will be decided on the technical merits of the corresponding processes. In this regard, metabolic engineering becomes of critical importance in advancing the competitiveness of biotechnology and realizing the vision of sustainable bioeconomy.

408788 Construction of Robust, Multi-Input Genetic Circuits and Memory Devices for Environmental Signal Sensing and Programmed Actuation

Abstract:

Signal sensing is essential for bacteria that live in fluctuating environments. However, even a group of isogenic bacteria, subject to identical culture conditions, often exhibit heterogeneity in phenotypes. This multi-stability can be explained by stochastic transitioning among multiple stable states of gene expression which are often the results of positive feedback loops in gene regulatory networks. Although such a "bet-hedging" strategy can increase bacterial fitness in unpredictable environments, such stochastic processes can prevent synthetic biologists from predictably engineering microbial cells that need to face uncontrolled environments. Thus, to fully exploit microbes in synthetic biology applications, complex gene regulatory systems and their behaviors in diverse conditions need to be better understood. To this end, we propose to construct "robust" genetic sensors and circuits that endow bacteria with the ability to sense environmental signals (e.g., temperature, light, pH, and oxygen level), perform computational operations, and implement programmed actuations (e.g., generation of toxins to kill parasite eggs).

By building sensors and circuits from the bottom-up, we have gained design principles for construction of programmable cells. Specifically, we have built robust logic gates that contain modular sensors for temperature or oxygen levels that can be more meaningful signals in the environment than chemical inducers that are used in the lab. In addition, we have constructed bistable switches that can generate noise-tolerant responses to fluctuating environmental inputs. These switches are based on the sequestering interaction between a transcription factor (an activator) and an anti-activator, along with multiple positive feedback loops. Importantly, these switches are functional over a wide range of parameters and endow the cells with a long-lasting memory. We are assembling these modular sensors, logic gates, and memory devices to construct robust, programmable probiotic strains that would reproduce in the intestine where parasite eggs are produced, and would come out of the human body with the parasite eggs and kill them. We will present progress towards development of such engineered microbes, which can be programmed to kill parasite eggs only when user-defined conditions are met.

409914 Synthetic Control of Transcription: A Decade's Perspective on Synthetic Promoter Design

Abstract:

Synthetic biology provides the means of designing novel parts to aid in the controlled, tunable, and regulated expression of individual genes, circuits, and pathways. Over the past decade, significant progress has been made in the area of promoter engineering. This talk will cover several advances in the area that range from random point-mutation guided approaches to new, model based design of synthetic parts. By utilizing such approaches, it is possible to create libraries of transcriptional control that can span several 100's of fold with respect to mRNA level. An emphasis will be placed on newer approaches to minimize the size of synthetic transcriptional control elements in yeast. Next, this talk will discuss the synthetic design of expression cassettes and operons and the impact of genetic context for synthetic parts—a particularly important aspect for characterization and predicable function. By synthetically altering upstream regions, terminator components, and promoter structure, defined and reproducible function can be obtained. Collectively, these results demonstrate novel approaches to synthetically alter and control gene transcription—a central goal of metabolic engineering and synthetic biology efforts. Once we have mastered these aspects, it is possible to design transcriptional control from scratch. This talk will conclude with progress toward this goal.

408893 Getting through the Gatekeepers: Changing the Selectivity of Semi-Permeable Cellular Membranes

Abstract:

The phospholipid bilayer has long been described as the cellular gatekeeper, preventing the passive transport of proteins, small molecules, and ions across cellular and organelle boundaries. It was recently discovered that some bacteria use organelle-like compartmental systems as well, contrary to the long-held belief that bacteria lack such organization, and these bacterial compartment membranes are made entirely of proteins. We explore the gatekeeping functions of each of these distinct types of membranes, studying protein and small molecule transport and sequestration in each case. Using various synthetic biology approaches, we engineer these systems in order to increase the yield of biochemical products such as butanol. Specifically, I will discuss how we can 1) sequester metabolic pathway enzymes and intermediates in the protein compartments and 2) gain control and alter the specificity of lipid membrane transporters for small molecules and proteins.

437446 Engineering Designer Biocatalysts and Pathways for Microbial Glycobiology

Abstract:

Carbohydrates add a level of diversity across all forms of life that is unparalleled by the information content of nucleic acids and proteins. The lack of a simple template to translate a glycan code into defined sugar structures contributes to this complexity and impedes efforts aimed at the production of biologically important glycans and glycoconjugates. With the discovery of glycoprotein synthesis in bacteria and functional transfer of glycosylation pathways between species, *Escherichia coli* cells have become a tractable host for understanding glycosylation and the underlying glycan code of living cells. Moreover, efforts to manipulate the pathways from sugar nucleotides to glycolipids to glycoproteins have transformed *E. coli* into a living factory for scalable, bottom-up production of complex glycoconjugates by design. Here, I will discuss our efforts to develop *E. coli* for the biosynthesis of a diverse array of glycan structures, which can be used to tailor the activity, stability, half-life, and shelf-life of a diverse array of protein targets ranging from biopharmaceuticals to industrial enzymes. I will also discuss our efforts to unify protein glycosylation in *E. coli* with the advanced techniques of protein engineering such as chimeragenesis, protein-fragment complementation, and directed evolution. The result is a powerful new platform for engineering the enzymes, pathways, end-products, and genomes of glycoengineered bacteria for creating the next generation of biotechnology-derived products.

435203 Metabolomic Modularity Analysis (MMA) to Quantify Human Liver Perfusion Dynamics

Tuesday, November 10, 2015: 3:15 PM
155A (Salt Palace Convention Center)

Gautham V. Sridharan[1], Bote Bruinsma[1], Martin L. Yarmush[1] and Korkut Uygun[2], (1)Center for Engineering in Medicine (CEM) at Massachusetts General Hospital – Harvard Medical School - Shriners Hospital for Children, Boston, MA, (2)Center for Engineering in Medicine, Harvard Medical School, Boston, MA

Biological cells are complex dynamic systems, comprising thousands of interacting nucleic acids, proteins, and metabolites acting in concert to perform coordinated and well regulated tasks. The ability to effectively quantify *system-wide* cellular responses to various stimuli is paramount to our understanding of complex disease-states and how pharmacological agents elicit desired outcomes. Investigators are therefore increasingly relying on large scale –omics data (transcriptomics, proteomics, metabolomics) to experimentally capture changes in signaling and metabolic pathway components. While the burden of data acquisition has lightened with improved instrumentation, data interpretation in the proper biological context remains challenging. Metabolomics data, in particular, are difficult to interpret in the context of pathways because a single metabolite may be utilized by several functional modules. Both univariate and multivariate statistical tools can determine which metabolites explain a differential response among experimental treatment groups. However, these data-driven methods do not utilize the vast domain-specific biochemical knowledge of known stoichiometric and regulatory interactions that can be used to systematically uncover which signaling or metabolic pathways are activated by a perturbation. To this end, mapping metabolomics data onto well curated metabolic graph networks can help identify interactions that may not be intuitive from simply observing the metabolic pathway maps offered by the KEGG database.

In this study, we introduce Metabolomic Modularity Analysis (MMA) as a graph-based algorithm to systematically identify modules of reactions enriched with metabolites flagged to be statistically significant based on univariate analysis. Briefly, a metabolic network is abstracted as a reaction-centric graph network where a non-directional edge is drawn between two reactions if a metabolite produced by the first reaction is consumed by the second. The length of the edge between reactions is weighted such that if the two reactions are involved in metabolites that are statistically significant, their edge distance is shorter. When applying Newman's hierarchical partition algorithm based on relative reaction-pair shortest paths, the resulting modules feature reactions from the surrounding local topological neighborhood, which favors adjacent reaction pairs that involve statistically significant metabolites. A defining feature of determining reaction-centric modularity is that interactions between reactions mediated by the production and consumption of cofactors and other hub metabolites are also accounted for.

We apply MMA on time-course metabolomics data collected from biopsies during subnormothermic machine perfusion (SNMP) of nine discarded human livers that were rejected for transplantation. These livers had endured various degrees of warm ischemia time (WIT) prior to organ procurement and three of them exhibited over 30% macrovesicular steatosis. Of the 155 primary metabolites measured, the pre-perfusion levels (t=0 hours) of 33 of them were found to be significantly correlated with WIT, suggesting that they could be putative biomarkers for ischemic injury. MMA was performed on the human ReconX metabolic network (7439 reactions and 2626 metabolites) after flagging these 33 metabolites in the cytosolic compartment as being significant. For a network this large, computing the adjacency distance matrix requires ${}_{7439}C_2$, or $2.77 \cdot 10^7$ computations for the initial network, which causes total run times on the order of days with a laptop, rendering the method impractical for repeated analysis. Fortunately, Mathworks® now offers Matlab's Distributed Computing Server (MDCS) on Amazon's Elastic Compute Cloud (EC2). For this study, one cluster node (c3.8xlarge) with 16 workers was rented at a rate of $1.68/hour for the cluster and $1.12/hour for MDCS. Several key parts of the MMA algorithm are parallelizable and was therefore able to complete a full run in 2.7 hours, with a total cost of $8.40 for the 3 hours of server time. The MMA partitioning resulted in 4755 hierarchical modules, of which 223 contain at least one significant metabolite. To highlight an example, one such module contained four metabolites significantly correlated with WIT; arachidic acid (arach[c]), cholesterol (chsterol[c]), stearic acid (ocdca[c]), and palmitic acid (hdca[c]). The stoichiometry of reactions in the module suggests that ischemia may have an impact on coenzyme A (coA[c]) pools as a cause for why these four metabolite levels are affected (Figure 1). Ongoing work involves applying MMA to identify modules activated by perfusion itself by flagging metabolites significantly different between various time points and determine how ischemia or fat content affects the modules uncovered.

In this study, we demonstrate that graph network-based analytics to quantify metabolomics data is made practical and feasible using parallel computing on Amazon's cloud. Prospectively, similar graph-based algorithms can be employed to analyze proteomics and transcriptomics data as well and ultimately devise ways to incorporate all information to best characterize cellular dynamics.

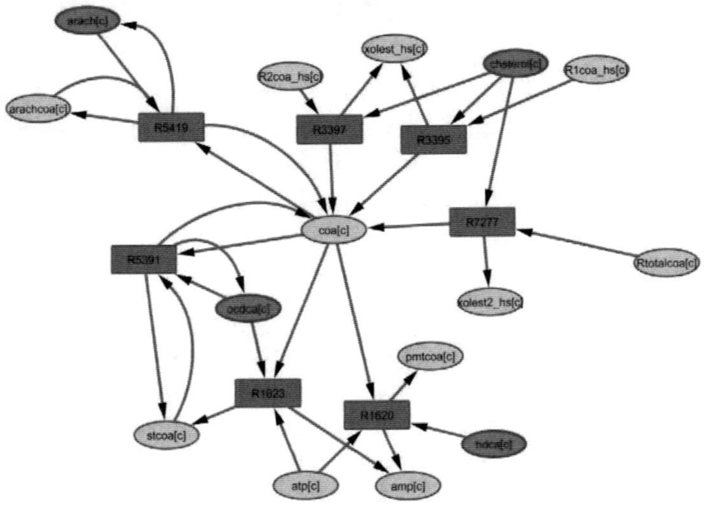

Figure 1 Example of module identified using MMA. Metabolite nodes are represented as ellipses, while reaction nodes are represented as rectangles. Metabolite nodes colored in red are those whose pre-perfusion levels are correlated to WIT (p<0.10).

419144 OM-FBA: Integrate Multi-Omics Data to Flux Balance Analysis to Better Understand Cell Metabolism

Abstract:

omFBA: Integrate Multi-omics Data to Flux Balance Analysis to Better Understand Cell Metabolism

Weihua Guo, Xueyang Feng*

Department of Biological Systems Engineering, Virginia Polytechnic Institute and State University, Blacksburg, VA 24060

Constraint-based metabolic modeling such as flux balance analysis (FBA) has been widely used to simulate cell metabolism. Thanks to its simplicity and flexibility, numerous algorithms have been developed based on FBA and successfully predicted the phenotypes of various biological systems. However, their phenotype predictions may not always be accurate in FBA because of using the objective function that is assumed for cell metabolism. To overcome this challenge, we have developed a novel algorithm, namely omFBA, to integrate multi-omics data (e.g. transcriptomics, proteomics and metabolomics) into FBA to obtain omics-guided objective functions with high accuracy. In general, we first collected multi-omics data and phenotype data from published database (e.g. GEO database) for different microorganisms such as *Saccharomyces cerevisiae*. We then developed a "Phenotype Match" algorithm to derive an objective function for FBA that can lead to the most accurate estimation of the known phenotype (e.g. biomass yield). The derived objective function was next correlated with multi-omics data via regression analysis to generate the omics-guided objective functions, which will be further used to accurately simulate cell metabolism. We have applied omFBA in both *E. coli* and *S. cerevisiae* and found that the ethanol yield and biomass yield can be accurately predicted in most of the cases tested (~70%) by using transcriptomics data. In addition, omFBA could be incorporated with other existing algorithms easily by using the omics-guided objective functions. Compared to current approaches that attempt to integrate omics data to FBA (e.g. ROOM), omFBA can be more advantages in extendibility and comparability.

422695 A Hidden Markov Model Reconstruction of a Cellular Signaling Network in Embryonic Fibroblasts from Time-Course Gene Expression Profiles Reveals the Mechanism of the SPRY2 Tumor-Suppressor

Abstract:

A hidden Markov model reconstruction of a cellular signaling network in embryonic fibroblasts from time-course gene expression profiles reveals the mechanism of the SPRY2 tumor-suppressor.

Mark F. Ciaccio and Neda Bagheri. Department of Chemical & Biological Engineering, McCormick School of Engineering, Northwestern University.

Introduction: Cellular phenotype is controlled, in part, by the abundance of signaling proteins as well as the dynamic parameters governing their activation. For this reason, time-course gene expression data is invaluable for understanding the molecular regulatory mechanisms that underlie cancer behavior. Sprouty Homolog 2 (SPRY2), a known tumor suppressor, not only attenuates the amplitude of Receptor Tyrosine Kinase (RTKs) signaling but also decreases the duration of activity [1]. However, the mechanism by which SPRY2 inhibits RTK-signaling is poorly understood. We performed RNA-sequencing at six time points on cell lysates after stimulation with Fibroblast Growth Factor (FGF) in Sprouty-null Mouse Embryonic Fibroblasts (MEFs) and wild-type controls. We trained a hidden Markov model on a transcription-factor/DNA binding database [2], and used it to predict the transcription factors responsible for bifurcation points where clusters of time-course transcriptional profiles diverge. The differential activity of transcription factors informed how the signaling network is rewired in the absence of SPRY2.
Background: SPRY2 is down-regulated in many types of cancers including those of the colon and breast [1].

Understanding the mechanism of SPRY2 in signal transduction can lead to new therapies for cancer. **Materials and Methods:** We quantified the expression of gene transcripts using RNA-seq at six time points in SPRY-knockout vs. wild-type MEF cells. A refined version of the Dynamic Regulatory Events Miner (DREM) algorithm [2] was used with a protein-DNA binding database to infer the most likely active transcription factors at each time point. Differences in edge confidences between signaling networks inferred from the changes in activity of transcription factors were used as the basis to infer the location of SPRY2 within the signaling network. **Results and Discussion:** We found SPRY2 increased binding of the ubiquitin ligase, CBL-B, to several RTKs including FGFR. This suggests that the Sprouty-proteins locate ubiquitin ligases to the cellular membrane, thereby marking the receptors for degradation. **Conclusions:** Hidden Markov models are an efficient means to infer the mechanism of proteins from time-course gene expression profiles. This study suggests the importance of the regulation of RTK-degradation in the etiology of cancer and highlights the importance of incorporating regulation of protein stability into computational models of cancer.

References:

[1] Edwin, Francis, et al. "Intermolecular interactions of Sprouty proteins and their implications in development and disease." *Mol pharm* 76.4 (2009): 679-691.

[2] Bar-Joseph, Ziv, et al.. "Studying and modelling dynamic biological processes using time-series gene expression data." *Nat Reviews Genet* 13.8 (2012): 552-564.

435867 Systematic Analysis of Adaptive Resistance and Fractional Responses of Melanoma Cancer Cells to RAF/MEK Inhibition

Abstract:

Activation of BRAF via a V600E mutation is the most prevalent genetic change in human melanoma (the deadliest form of skin cencaer), found in >50% of tumors, and induces constitutive activation of pro-mitogenic RAF/MEK/ERK signaling. Treatment of $BRAF^{V600E}$ melanomas with drugs, such as vemurafenib, that inhibit RAF/MEK signaling is effective in the short term, but remission is not durable. Drug resistance appears to involve short-term adaptive responses that compensate for RAF/MEK inhibition via up-regulation of other pro-growth mechanisms. Understanding and ultimately preventing adaptive responses is key to durable therapy. Systematic data comparing $BRAF^{V600E}$ tumor cells is generally lacking and it is not known whether adaptation is fundamentally similar across cell types or among individual cells within a cell population.

We apply a systematic approach to studying the responses of human melanoma cell lines to five RAF and MEK inhibitors, with the overall goal of (i) characterizing variability in adaptation with time, dose, cell type and across individual cells, (ii) discovering new or poorly characterized adaptive mechanisms, and (iii) demonstrating the effectiveness of a high-throughput approach involving multiplex measurement, single-cell analysis and computational modeling. The data involves time-course measurement of total level and activity of signaling proteins and cell state markers using array-based methods and single-cell immunofluorescence assays as well as measurement of apoptosis and cell viability under the same conditions. Statistical modeling using partial least squares regression (PLSR) revealed which of the changes in the ~200,000 point dataset were phenotypically consequential.

We found that responses to RAF inhibitors are remarkably diverse and involve multiple pathways that can be up or down-regulated over time, with significant variability across cell types and individual cells. We identified a role for JNK/c-Jun signaling in altering the cell-cycle distribution of melanoma cells, causing apoptosis-resistant cells to accumulate and drug maximal effect (Emax) to fall; co-drugging with RAF and JNK inhibitors or JUN knockdown reverse this effect. Our study shows that a systems-level approach (combining high density time-dependent measurements, quantitative modeling and single-cell analysis) may provide a general framework for evaluating new drugs with adaptive and paradoxical response, and identifying potentially useful combination therapies.

434054 Synechocystis Sp. PCC 6803: Light Plus Endogenous Regulation Governing Gene Expression Patterns in the Diurnal Cycle

Abstract:

Cyanobacteria are the only photosynthetic prokaryotes with circadian behavior. Based on global transcriptomic data, several cyanobacteria including the strain *Synechocystis* sp. PCC 6803 have been analyzed thus revealing their oscillatory temporal expression. *Synechocystis* cells tune the transcriptional level of genes in response to the oscillation of daily diurnal cycle. However, it is still unclear how changes in cellular physiology are coordinated by this cyclic light exposure. In this study, we measured and subsequently analyzed expression patterns of nearly 3,500 genes in *Synechocystis* 6803 over two consecutive diurnal periods. Based on this analysis, at least 39% of genes with many from major cellular processes including central carbon metabolism, transport, and regulation were found to exhibit oscillating expression profiles. The pattern of gene expression also led to the development of two distinct transcriptional networks of co-regulated oscillatory genes. These networks described how *Synechocystis* regulated its cellular processes at the end of the dark period so as to make efficient use of light at the early light period. Interestingly, the light peaking genes showed damping patterns even in the middle of the light period which led to the inference not only the presence/absence of light but also endogenous regulation controlled the oscillatory gene expression patterns. We then measured important physiological parameters including glycogen concentration, ATP level, and $NADPH/NADP^+$ ratio in two consecutive light/dark cycles and, thereby, analyzed the physiological changes and their correlations with temporal expression patterns. This global transcriptional analysis of *Synechocystis* sheds light on the global coordination of cellular processes as evidenced in the transcriptional levels of genes.

429940 Applying Adaptive Laboratory Evolution to Engineer Platform Strains

Abstract:

Adaptive Laboratory Evolution (ALE) is a powerful technology for developing both compound-specific and platform bioprocessing strains. To leverage the utility of high-throughput sequencing after ALE, sophisticated experimentation is necessary to perform controlled high-throughput ALE experiments with the goal of understanding the range of phenotypes available to cellular production systems and how they arise. The presented work will describe an automated technology built for performing ALE experiments and its application towards highlighted use cases. These ALE use cases include adapting strains to higher tolerance of a desired compound, adaptation to utilize non-natural substrates, and adaptation to various bioprocessing conditions. The genomic changes which enable such adaptation will also be discussed along with interpretation of optimal phenotypes using genome-scale modeling and omics data characterization. ALE use is increasing in industrial biotechnology and sophisticated platforms for conducting experiments and analyzing outcomes will be necessary to make it a viable technology in the strain generation pipeline.

412226 Genome Editing in *Yarrowia Lipolytica* Using a Crispr-Cas9 System with RNA Polymerase II Transcribed Single Guide RNA

Abstract:

Yarrowia lipolytica has been the focus of a wide range of studies due to its ability to make and accumulate lipids to a high level. Here, we engineer a single plasmid CRISPR-Cas9 system for *Y. lipolytica* that enables markerless gene knockouts and insertions. A functional system was created using a codon optimized Cas9 from *Streptococcus pyogenes* and single guide RNA (sgRNA) expressed from the well characterized constitutive TEF promoter. RNA cleavage by flanking ribozymes generated the mature sgRNA, ensuring proper folding and preventing export to the cytosol by mRNA transport processes. Gene knockout efficiency of ≥75% was achieved when targeting the *ku70* gene without a cotransformed homologous recombination (HR) donor. The *ku70* gene was targeted as it has been shown that *Y. lipolytica* strongly favors non-homologous end-joining (NHEJ) to repair double strand breaks. Knockout of *ku70* suppresses NHEJ and allows HR to serve as the primary double strand break repair mechanism. Using the generated *Δku70* strain and the CRISPR-Cas9 system, markerless gene tagging with fluorescent proteins is being done to investigate protein localization to lipid droplets and to the endoplasmic reticulum.

437698 The Emerging Organism Engineering Industry

Abstract:

There is an emerging demand for sourcing plant-derived extracts (nutraceuticals, flavors, fragrances, sweeteners, etc.) from engineered microbes. While recent advances in synthetic biology and metabolic engineering provide feasible approaches to engineering such organisms, commercial success for developing these "cultured" ingredients presents specific challenges. Unlike biofuels, where efforts can be focused on one particular molecule given the enormous market size, cultured ingredients require developing different organism lines in a rapid and low cost fashion. This requires a scalable solution for bio-manufacturing of organisms, which is provided by our state of the art foundry that continues to grow. I will describe how organism development at Ginkgo leverages our foundry to accelerate the design/build/test using specific examples. In particular, I will highlight the value of combining computer-aided engineering software tools, cheap gene synthesis and high resolution-accurate mass LCMS to develop engineered microbes. Finally, I'll touch on how our improvements in manufacturing organisms lend to opportunities outside of cultured ingredients.

437697 Many Microbes, Many Molecules; Using Automation and Big Data to Create Better Bugs for Industrial Fermentation

Abstract:

Zymergen has built an automated, high throughput strain engineering platform that is flexible enough to work with diverse micro-organisms, including those used at scale in the largest fermented products markets. Throughput allows us to test thousands of design ideas across many library types to find the ideas that improve performance. Automation makes our strain improvement process precise and repeatable, which means we can identify ideas that increase performance by only a few percent and find the next set of good ideas in the same time it took to find the first. As we reliably build and test design ideas across diverse microbes and pathways, we create a rich data set that we use to refine our approach. This data informs not only the library types we use to solve a specific problem, but also the protocol and process improvements that will have maximal benefit.

436770 Commercial Production of Omega-3 Eicosapentaenoic Acid By Metabolic Engineering of an Oleaginous Yeast, Yarrowia Lipolytica

Abstract:

The omega-3 fatty acids, cis-5, 8,11,14,17-eicosapentaenoic acid (C20:5; EPA) and cis-4,7,10,13,16,19-docosahexaenoic acid (C22:6; DHA), have wide-range health benefits for human and animals. Currently, the major source for EPA and DHA is from fish oil, and a minor source of DHA is from microalgae. With the increased demand for EPA and DHA, DuPont has developed a clean and sustainable EPA source through fermentation production using metabolically engineered strains of *Yarrowia lipolytica*. Specifically, EPA biosynthetic and supporting pathways have been introduced into this oleaginous yeast to synthesize and accumulate EPA under fermentation conditions. The engineered *Y. lipolytica* strains have been used for commercial production of EPA-rich oil and EPA-rich biomass by fermentation. Two commercial products have been developed, "New Harvest", a clean, renewable and vegetarian oil with EPA at 55% (weight/weight) for human nutrition supplement and "Verlasso[R]", a premium branded salmon which is sustainably farmed due to incorporating EPA-rich *Yarrowia* biomass in the feed. The yeast triacylglyceride oil has a unique fatty acid profile with less than 5% as saturated fatty acids and more than 55% as EPA. The oil with this fatty acid profile provides significant health benefits for human and animals. A most advanced *Yarrowia* production strain has been generated to produce EPA at >28% dry cell weight, a titer that is more than 90% higher than was obtained with the strain we previously reported (Nature Biotechnology, 2013, 8: 734-740). Our land-based production of omega-3 and omega-6 fatty acids with tailored compositions provides a superior source for these high value molecules for applications in human nutritional supplements, functional foods, infant foods, medical foods, pharmaceuticals, and animal feeds.

437583 Recreating Natural Plant Processes in Bacteria for the Production of Complex Biomolecules

Abstract:

Despite the potential utility of specialized secondary metabolites in a number of commercial applications (pharmaceuticals, consumer and food products, and industrial chemicals), their low abundance in nature presents a significant sourcing and economic obstacle for adoption. Here, we describe the advantages of pairing multivariate modular metabolic engineering of bacteria with systems-level characterization of engineered strains by transcriptomics, proteomics, and metabolomics for expediting the development of economical and sustainable processes for the production of these complex biomolecules.

437691 Industrial Cell-Free Bioprocessing

Abstract:

GreenLight Biosciences has developed a flexible bio-manufacturing platform that leverages proprietary breakthroughs in Cell-free Bioprocessing. Cell-free Bioprocessing takes full advantage of the high selectivity, yield, and sustainability of biological catalysts without the burden of maintaining a living organism under more rigid physiological conditions. Bioprocess development is accelerated through the rapid and precise testing of biocatalytic system variation without the need to reengineer organisms. This revolution in bio-manufacturing has allowed GreenLight to create a new generation of bioprocesses capable of addressing major issues of sustainability and feedstock flexibility at costs, yields, and productivities that compete favorably with existing chemical processes.

437696 Successful Application of Computational Tools to the Design of Industrial Enzymes and Metabolic Pathways for the Production of Novel Molecules and Materials

Abstract:

The recent emergence of industrial and synthetic biotechnology has the potential to radically transform how we synthesize and produce chemicals. As explained by E.J. Corey's Nobel in Chemistry lecture, one of the main driving forces behind the great variety of compounds manufactured by chemists since the 50s has been advances in synthetic chemistry approaches, especially retrosynthetic techniques that starting from a desired compound propose multiple synthetic pathways from known intermediates. Rapidly investigating a large variety of retrosynthetic pathways is fundamental to synthetic biology. To this end, Arzeda has developed a computational pathway prospecting tool that can enumerate all possible enzyme pathways from a given starting molecule to a compound of interest- using existing and novel enzyme activities. In addition to and complementary to the pathway tool, Arzeda's proprietary technology includes novel computational methods to rationally engineer enzymes with a wide range of activities using known catalytic mechanisms for non-native substrates all the way to de novo catalytic machineries. Taken together Arzeda's computational techniques enable elegant synthetic biology solution to some of the most difficult small molecule synthesis problems.

415742 Chemical-Genetic Inference of Antibiotic Interactions for Combination Therapies

Abstract:

Combination antibiotic therapies are being increasingly used in response to a global rise in drug resistance. Mapping synergistic and antagonistic interactions between antibiotics is essential for designing effective combinations that enhance potency and counter resistance. However, the large search-space of candidate drugs and dosage regimes makes such approaches highly challenging. Here, we present a computational approach that integrates publicly available chemogenomic data to predict dose-specific antibiotic interactions. We experimentally tested interactions for 171 antibiotic pairs in nine dose combinations in *Escherichia coli* to validate our approach. Our analysis revealed a core set of cellular pathways (e.g., central metabolism) and biophysical factors (e.g., drug lipophilicity) that are predictive of synergy and antagonism. Using an evolutionary approach, we modified our chemical-genetic model of *E. coli*, a Gram-negative bacterium, to accurately predict drug-drug interactions in the Gram-positive pathogen *Staphylococcus aureus*. This study provides a framework for chemogenomics-driven discovery and development of effective combination therapies.

431903 Dissecting the Effects of (5Z)-4-Bromo-5-(bromomethylene)-3-Butyl-2-(5H)-Furanone on Bacillus Anthracis Metabolism

Abstract:

Quorum sensing inhibitors or quorum quenchers hold great potential as therapeutics for treating infectious diseases. One such molecule, *(5Z)*-4-bromo-5-(bromomethylene)-3-butyl-2-*(5H)*-furanone (furanone) has been found to inhibit the quorum sensing circuit of *Bacillus anthracis*, the etiological agent of anthrax. However, the impact of furanone on the cellular system as a whole remains unclear. In this work, the affect of furanone on cellular metabolism was evaluated through an integrated analysis involving genome-scale metabolic modeling and experimental efforts. Specifically, a metabolic model of *B. anthracis* consisting of 1,003 metabolites and 1,049 reactions was used to analyze experimental results collected for the Sterne strain of *B. anthracis* when grown in the presence of furanone. It was found that although growth rate and glucose uptake rate both decreased in the presence of furanone, energy demand increased. It was postulated that the higher energy requirement was necessary to accommodate critical energy-dependent translocation processes of key nutrients. Additionally, the introduction of furanone appeared to alter the redox potential generation pathway, indicating that regeneration of NAD+ occurred by electron transfer to glycerol-3-phosphate rather than fermentation. The overall impact appeared to be indicative of an oxidative stress response.

422573 Discovery and Dissection of Metabolic Oscillations in the Nitric Oxide Response of *Escherichia coli* Under Microaerobiosis

Abstract:

Nitric oxide (NO•) is an important antimicrobial produced by the immune system to impair pathogens. Many pathogens, such as *Mycobacterium tuberculosis*, *Salmonella enterica*, and pathogenic *Escherichia coli*, harbor NO• detoxification and repair systems which have been linked to their virulence, and thus present an attractive target toward which novel anti-infective therapies could be designed. Unfortunately, the few compounds that have been found to inhibit major NO• detoxification systems in bacteria suffer from poor intracellular transport or severe toxicity toward humans, warranting a search for additional targets in bacterial NO• defense systems. The broad reactivity of NO• and its reaction products give rise to an expansive, complex reaction network, necessitating the use of a computational approach to effectively study the system in its entirety. To facilitate quantitative exploration of the *E. coli* NO• stress response, we previously constructed a kinetic model of the intracellular NO• biochemical network, including relevant processes such as enzymatic NO• detoxification, iron-sulfur cluster nitrosylation and repair, DNA deamination and repair, transcriptional regulation, and NO• autoxidation. Model simulations were validated with experimental measurements of [NO•], [NO_2^-], [NO_3^-], and [O_2] in *E. coli* cultures, and facilitated the discovery of a strong decrease in utility of the main aerobic NO• detoxification enzyme, NO• dioxygenase (Hmp), with increasing NO• delivery rate. Furthermore, Hmp was predicted to dominate cellular NO• consumption even into the microaerobic regime (35 μM dissolved O_2), which was confirmed experimentally.

Given that O_2 levels at *E. coli* infection sites *in vivo* range from anaerobic to microaerobic, we used the NO• model to investigate the dynamics of the *E. coli* NO• response network within this medically-important regime. Although the two major NO• detoxification systems in *E. coli*, Hmp and NorV, are known to dominate NO• consumption under aerobic and anaerobic conditions, respectively, little is known about their participation and dynamics within the microaerobic regime that lies between these two extremes. The model predicted a deficiency in the overall rate of NO• detoxification for O_2 environments between anaerobic (0 μM O_2) and the upper-end of microaerobic (50 μM O_2), which was attributed to an insufficient overlap of Hmp and NorV activity. Experimental measurements of NO• detoxification at these intermediate O_2 concentrations (5, 10, and 20 μM O_2) confirmed the predicted trend; however, the extent to which NO• detoxification was impaired was underestimated, and oscillations appeared in the [NO•] curves measured at 5 and 10 μM O_2. Efforts to reconcile the predictions with experimental measurements yielded a model that could quantitatively capture NO• dynamics throughout the microaerobic regime. An investigation of model parameters contributing to the oscillations predicted a mechanism involving competition between Hmp and respiratory cytochrome oxidases for O_2, which was confirmed experimentally through model-guided biochemical and genetic analyses. This demonstrates the utility of a modeling approach to quantitatively study the bacterial NO• defense network even under previously uncharacterized environmental conditions, and how it can facilitate the dissection of complex phenomena to yield novel insight about the quantitative interplay between network components.

432307 Identification of Gut Microbiota-Derived Metabolites As Liver Inflammation Modulators in Non-Alcoholic Fatty Liver Disease

Abstract:

Identification of Gut Microbiota-derived Metabolites as Liver Inflammation Modulators in Non-alcoholic Fatty Liver Disease

The human gastrointestinal (GI) tract harbors an estimated 10 to 100 trillion microorganisms, encompassing several hundred different species. Collectively referred to as the gut microbiota, the microbes perform a number of essential physiological functions, including the metabolism of complex carbohydrates [1], development of the immune system [2], and defense against pathogens [3]. Beyond the GI tract, gut-brain, gut-lung, and gut-liver associations have also been identified, highlighting the importance of the microbiota in host physiology. Consistent with the view that the microbiota is critical for whole body homeostasis, alterations in the intestinal microbiota composition (i.e. dysbiosis) have been correlated to various diseases, including obesity [4, 5], diabetes [6], and cancer [7].

Non-alcoholic fatty liver disease (NAFLD) has emerged as the most prevalent chronic liver disease in Western countries, paralleling the rise in obesity and the metabolic syndrome. It is characterized by steatosis, or fat deposition in the liver, which can remain a benign condition. However, in 10-20% of patients, NAFLD progresses from steatosis to non-alcoholic steatohepatitis (NASH), which can lead to cirrhosis and liver cancer. Seminal work by the Gordon laboratory [8, 9] and other groups demonstrated that dysbiosis could be a contributing mechanism for obesity and related diseases, including NAFLD, whose estimated prevalence in obese individuals exceeds 75% [10]. Direct, potentially causal involvement of the microbiota in NASH was highlighted by Le Roy et al. [11], who reported that high fat diet (HFD)-fed germ-free (GF) mice inoculated with bacterial isolates from feces of HFD-fed hyperglycemic mice developed steatohepatitis, whereas HFD-fed GF mice inoculated with bacteria from HFD-fed, but normoglycemic mice only showed mild steatosis. This suggests that the composition of the intestinal microbiota, and hence their metabolic products, could determine whether NAFLD progresses to NASH. While these and other recent studies have begun to build a strong case for potential mechanistic links between dysbiosis and NASH, the identities of microbiota-derived molecular mediators and their target pathways remain to be fully elucidated. The study described in this abstract utilizes both untargeted and targeted metabolomics to identify and characterize microbiota-dependent immunomodulatory metabolites that act on the liver to modulate liver inflammation in steatosis.

In previous work [13], we identified a panel of aromatic amino acid (AAA)-derived metabolites whose levels in the mouse intestine depended on the presence of the microbiota. Furthermore, several of these metabolites were found to activate the aryl hydrocarbon receptor (AhR), a transcriptional regulator of host-microbiota interactions in the intestine. To investigate a potential role for these metabolites in NAFLD, we compared their profiles in serum and liver samples from 14-week old male C57BL/6J mice raised for 8 weeks either on HFD or low-fat diet (LFD). The diets were sucrose calorie-matched and derived 60% and 10% of calories from fat, respectively. Metabolites were extracted from cold PBS-rinsed liver tissue and serum samples using a solvent-based method. Using an untargeted method, we identified 73 AAA-derived metabolites, including host- and microbiota-dependent metabolites as well as their methyl and hydroxyl derivatives. Interestingly, a large fraction of metabolites that were significantly elevated or depleted in liver and serum from HFD-fed mice (compared to LFD-fed mice) were the hydroxyl/methyl derivatives. Several of the metabolites that were consistently detected in all four sets of samples (liver and serum, HFD and LFD) were previously found to depend on the microbiota and activate the AHR [13]. We thus selected a pair of representative metabolites from this set, tryptamine (TA) and indole-3-acetate (I3A), for further analysis.

Absolute quantification of TA and I3A using targeted analysis confirmed that their levels were significantly elevated (ca. 2- and 3-fold, respectively) in the serum of LFD mice compared to HFD mice. We then investigated the ability of TA and I3A to modulate inflammation in hepatocytes using both human (HepG2) and murine (AML12) cell lines. Cultured HepG2 cells were treated with 1, 10, or 100 mM TA or I3A for 24 h, and then exposed to the pro-inflammatory cytokine TNFα for another 24 h in the presence of the metabolites. The expression of two anti-inflammatory genes (IL-4 and IL-10) and a pro-inflammatory chemokine (MCP1) were determined using qPCR. For all three markers, pretreatment with I3A or TA at the 10 mM dose significantly attenuated the pro-inflammatory responses elicited by TNFα treatment. These trends were in good agreement with the results obtained using the AML12 murine cell line.

Using HepG2 cells, we next investigated whether the AAA-derived, microbiota-dependent metabolites could modulate the impact of TNFα treatment on cholesterol and bile acid metabolism. Compared to vehicle control, TNFα treatment significantly increased both intra- and extracellular levels of several bile acids, including cholic acid (CA), glycocholic acid (GCA), and taurocholic acid (TCA). This profile was consistent with our findings that both CA and TCA were significantly elevated in livers of HFD-fed mice compared to LFD-fed mice. Pretreatment with either 1 mM TA or I3A significantly attenuated the TNFα–stimulated increases in bile acid levels, similar to the metabolites' effects on cytokine and chemokine gene expression.

To confirm activation of the AHR pathway, we utilized a hepatocyte reporter cell line, where binding of an AHR-ligand complex to the dioxin response element (DRE) leads to transcription of the reporter gene (GFP), resulting in fluorescence. Consistent with our previous findings [13], 24 h exposure to TA and I3A dose-dependently increased GFP expression in the reporter cells, indicating ligand activation of the AHR by these metabolites.

In summary, our results demonstrate that several AAA-derived microbiota metabolites are significantly depleted in a mouse model of diet-induced liver steatosis, and that these metabolites can act directly on hepatocytes to modulate inflammatory pathways. Our results also show that these microbiota metabolites are ligands for the AHR, providing a mechanistic link for the observed anti-inflammatory effects. Taken together, our findings support the hypothesis that dysbiosis of the gut microbiota could predispose the liver to inflammation in diet-induced steatosis through an altered microbiota metabolite profile. Prospectively, insights into the mechanisms underlying the link between microbiota dysbiosis and NAFLD could provide novel strategies to treat or prevent the progression of fatty liver diseases through the use of probiotics or postbiotics.

[1] T. Arora and R. Sharma, "Fermentation potential of the gut microbiome: implications for energy homeostasis and weight management," *Nutrition Reviews,* vol. 69, pp. 99-106, 2011.

[2] L. V. Hooper, D. R. Littman, and A. J. Macpherson, "Interactions Between the Microbiota and the Immune System," *Science,* vol. 336, pp. 1268-1273, June 8, 2012 2012.

[3] N. Kamada, G. Y. Chen, N. Inohara, and G. Nunez, "Control of pathogens and pathobionts by the gut microbiota," *Nat Immunol,* vol. 14, pp. 685-690, 2013.

[4] J. Shen, M. S. Obin, and L. Zhao, "The gut microbiota, obesity and insulin resistance," *Molecular Aspects of Medicine,* vol. 34, pp. 39-58, 2013.

[5] S. Greenblum, P. J. Turnbaugh, and E. Borenstein, "Metagenomic systems biology of the human gut microbiome reveals topological shifts associated with obesity and inflammatory bowel disease," *Proceedings of the National Academy of Sciences,* vol. 109, pp. 594-599, January 10, 2012 2012.

[6] J. K. Nicholson, E. Holmes, J. Kinross, R. Burcelin, G. Gibson, W. Jia, *et al.*, "Host-Gut Microbiota Metabolic Interactions," *Science,* vol. 336, pp. 1262-1267, June 8, 2012 2012.

[7] J. C. Arthur, E. Perez-Chanona, M. Mühlbauer, S. Tomkovich, J. M. Uronis, T.-J. Fan, *et al.*, "Intestinal Inflammation Targets Cancer-Inducing Activity of the Microbiota," *Science,* vol. 338, pp. 120-123, October 5, 2012 2012.

[8] R. E. Ley, F. Backhed, P. Turnbaugh, C. A. Lozupone, R. D. Knight, and J. I. Gordon, "Obesity alters gut microbial ecology," *Proc Natl Acad Sci U S A,* vol. 102, pp. 11070-5, Aug 2 2005.

[9] P. J. Turnbaugh, R. E. Ley, M. A. Mahowald, V. Magrini, E. R. Mardis, and J. I. Gordon, "An obesity-associated gut microbiome with increased capacity for energy harvest," *Nature,* vol. 444, pp. 1027-31, Dec 21 2006.

[10] S. G. Sheth, F. D. Gordon, and S. Chopra, "Nonalcoholic steatohepatitis," *Ann Intern Med,* vol. 126, pp. 137-45, Jan 15 1997.

[11] T. Le Roy, M. Llopis, P. Lepage, A. Bruneau, S. Rabot, C. Bevilacqua, *et al.*, "Intestinal microbiota determines development of non-alcoholic fatty liver disease in mice," *Gut,* vol. 62, pp. 1787-94, Dec 2013.

[12] W. R. Wikoff, A. T. Anfora, J. Liu, P. G. Schultz, S. A. Lesley, E. C. Peters, *et al.*, "Metabolomics analysis reveals large effects of gut microflora on mammalian blood metabolites," *Proceedings of the National Academy of Sciences of the United States of America,* vol. 106, pp. 3698-3703, Mar 2009.

[13] G. V. Sridharan, K. Choi, C. Klemashevich, C. Wu, D. Prabakaran, L. B. Pan, *et al.*, "Prediction and quantification of bioactive microbiota metabolites in the mouse gut," *Nat Commun,* vol. 5, p. 5492, 2014.

429442 Identifying Motifs Associated with Autoimmune Diseases Using Next Generation Sequencing Experiments

Abstract:

Autoimmune diseases are typically chronic, have major quality-of-life effects and afflict a large minority of the population. A hallmark of autoimmune diseases is the presence of antibodies that bind to "self" targets. Unfortunately, it has proven challenging to identify these autoantibodies and develop accurate diagnostic tests for their presence in many diseases.

We developed a method to Identify Motifs Using Next generation sequencing Experiments (the IMUNE method) capable of discovering disease-associated motifs. First, a large, bacterial-display peptide library is screened using magnetic and/or fluorescence-activated cell sorting to recover peptides bound by an individual's antibodies. Next generation sequencing technologies (e.g. Illumina NextSeq) are utilized to identify these millions of peptides. Because of the large diversities in the peptide library and in patients' antibodies, it is unlikely that this experimental approach alone will identify peptides bound by many disease patients but not control patients. Therefore, computational analysis is employed to identify patterns of amino acids that are statistically enriched in a cohort of patients compared to control patients, where a pattern is a series of specific amino acids possibly interspersed with undefined residues. Finally, these enriched patterns are clustered together to create the disease-associated motifs. We will present an overview of the experimental and computational methods, and show examples from Celiac disease and multiple sclerosis to highlight the method's ability to discover known and novel disease-associated motifs.

431408 Novel Approach to Target Endogenous KRAS Mutations By the Crispr/Cas9 Technology

Abstract:

CRISPR/Cas9 is an enabling RNA-guided technology for genome targeting and engineering. Here we demonstrate that the lentiviral-mediated CRISPR/Cas9 technology can be used to specifically target single-nucleotide heterozygous mutations while exerting no aberrant effects on the wild-type alleles. Specifically, we used a novel gRNA-constraint-mediated method to target the heterozygous G13A activating mutation of KRAS in colorectal cancer cells and we show reversal of drug resistance to a MEK small-molecule inhibitor. Our study introduces a new paradigm in genome editing and therapeutic targeting via the use of gRNA to guide Cas9.

433019 Rewiring Cellular Input-Output to Engineer Cell-Based Therapies That Interface Robustly with Human Physiology

Abstract:

Cell-based therapies have proven to be useful for treating a wide variety of diseases including autoimmune disease, infectious diseases, and cancer. A particularly promising frontier is the use of engineered cell therapies, wherein cells are programmed to carry out custom functions such as harnessing the immune system to find and destroy cancerous cells. Such therapies have now achieved robust clinical successes for some applications. However, controlling or modulating the activity of these therapies post-implantation remains both attractive, to enhance both safety and therapeutic efficacy, and challenging, since the tools necessary to engineer custom functional programs are limited. In particular, engineering cells to sense and recognize specific combinations of environmental cues would be desirable, but no such technology has yet been reported. Toward this goal, our lab has developed a platform for engineering novel protein biosensors, termed modular extracellular sensor architecture (MESA), for detecting exclusively extracellular cues. Upon binding extracellular ligand, MESA receptors release sequestered transcription factors from the cytoplasmic face of the plasma membrane, freeing the factor to regulate expression of an "output" gene or genes within the cell. Here we present three areas in which we have recently expanded upon the capabilities of the core MESA technology: (1) achieving novel ligand recognition by integrating modular ligand-binding domains, (2) integrating MESA with intracellular gene circuits that enable the cell to "process" sensory information by logical evaluation, and (3) rewiring cellular input-output by coupling extracellular sensing to intracellular modulation of endogenous gene expression.

First, as modular ligand binding domains, we used camelid single domain antibody fragments termed "nanobodies". Each nanobody's small size enables multiple nanobodies to recognize distinct, non-overlapping epitopes on a ligand of interest. Thus nanobody MESA are able to recognize monomeric or asymmetric ligand inputs. After demonstrating the feasibility of nanobody MESA, we integrated multiple such receptors into a logical processing circuit that enables the multiplexed logical evaluation of distinct protein ligand "inputs". Next, we demonstrated that human antibody-derived scFv could also be used as ligand-binding domains in MESA, and VEGF-MESA were thus designed to sense the tumor-associated cytokine, VEGF. In order to regulate endogenous genes, the MESA transcription factor domain was replaced with the dCas9 protein, which is able to act as a 'pseudo-transcription factor' by tethering transactivation domains to the dCas9 protein and providing small guide RNAs in trans to direct the Cas9 protein to a specific genomic locus. As proof of principle, we developed VEGF-MESA that induce expression of IL-2, a cytokine important for T-cell growth and proliferation. In principle, the dCas9 protein can be guided to any genomic locus of interest by simply delivering a different small guide RNA. Cells expressing VEGF-MESA exhibited minimal background signaling in the absence of ligand, and such cells showed significant increases in IL-2 gene expression upon exposure to ligand. Thus, this initial study demonstrates that MESA may be harnessed to effectively rewire cellular input-output behavior. This work lays the foundation for leveraging MESA receptors to boost the safety and efficacy of cancer immunotherapies as well as cell-based therapies for a range of clinical applications.

436637 Understanding Tumor Microenvironment's Metabolic Regulation of Cancer Cells

Abstract:

Tumor cells need appropriate nutrient for growth, invasion and energy metabolism. A growing number of studies indicate that metabolic dependencies of cancer cells are regulated by **tumor microenvironment (TME)** which rewires the cellular metabolism to promote tumorigenicity. Glutamine can play a critical role in cellular growth in multiple cancers. Recently, we uncovered a missing link between cancer invasiveness and glutamine dependence. Using isotope tracer and bioenergetic analysis, we showed that low invasive ovarian cancer (OVCA) cells are glutamine independent, whereas high invasive OVCA cells are markedly glutamine dependent. Here, we present a multidimensional approach, where we synergistically integrate *in vitro* stable isotope metabolomics, cell functional assays (invasion, migration, proliferation, 3D cocultures), mitochondrial bioenergetics, clinicopathological features (patient derived CAFs and stromal fibroblasts, and patient derived cancer cells), targeted therapies using in culture and *in vivo* models, to understand TME regulation of OVCA growth and metastasis. We will demonstrate a novel mechanism by which stromal cancer associated fibroblasts cells maintain tumor growth in OVCA cells.

AUTHOR INDEX

AUTHOR INDEX

AIChE
120 Wall Street, FL 23
New York, NY 10005-4020

ISBN 978-1-5108-1847-7

CURRAN ASSOCIATES INC.
proceedings.com